Making of a Satellite Centre

Prem Shanker Goel

# Making of a Satellite Centre

The Genesis of ISRO's URSC

Prem Shanker Goel
National Institute of Advanced Studies
INAE Satish Dhawan Chair of Engineering
Eminence
Bengaluru, Karnataka, India

ISBN 978-981-16-3482-6      ISBN 978-981-16-3480-2   (eBook)
https://doi.org/10.1007/978-981-16-3480-2

This Springer imprint is published by the registered company Springer Nature Singapore Pte Ltd.
The registered company address is: 152 Beach Road, #21-01/04 Gateway East, Singapore 189721, Singapore

# Prolog

Indian Space Research Organisation (ISRO) is a unique organisation. It has a culture which is unique, and there is no other way to describe it except to say it is the ISRO culture. The Satellite Program of ISRO has done wonderful things, like providing confidence to the country with the launch of Aryabhata satellite on 19 April 1975, that we can do complex technology, when even safety pins were being imported into the country. It was a simple satellite by today's standard, but a big technology leap for the country in the seventies. Pokaran-1 in 1974 and Aryabhata in 1975 were two technology events that gave Indian scientists and engineers confidence to venture into new areas, which were not so far pursued in India.

The satellite technology growth has been slow but consistent, towards complexity, modernisation and professionalism. While URSC has built 104 satellites (as on Dec 31, 2019), a few missions have been landmark missions that opened the pathway for the future. Aryabhata, Bhaskara-1, APPLE, IRS-1A, INSAT-2A, IRS-P3, MetSat (Kalpana), INSAT-2E, CARTO-1, SRE, TES and IRNSS are some of those landmarks. Each one raised the bar in technology development and mission management. Chandrayaan-1, which attracted the imagination of the country, was another great achievement of mission management. Mangalyaan further raised the capability to manage a mission beyond the earth's gravity. Now we have a robust technology base and operational space systems for Remote Sensing, Communication and space-based navigation and space science. Chandrayaan-2 is yet another advanced technology and mission management demonstration. Though the Lander Vikram had hard landing very near to the pre-designated spot on the Moon, it was a remarkable mission in technology development and mission management. Gaganyaan (The Manned Mission) on its way is another technology challenge and dream of an aspiring nation.

This book is the story of people behind satellites, narrated through Satellite Centre (ISAC or URSC), which represents satellite activity as a living organism. It is the story of remarkable people, a few named and many unnamed, who created and nurtured this body, which I believe has conscious of its own. Like any other living body, it has gone through its ups and downs but has always bounced back with full energy after every hiccup.

As this story is of people, narrated by an individual who was one amongst them, it is likely to be biased in impressions and interpretations. I have, however, tried to be as fair as possible towards my colleagues, each one of whom has contributed to the growth of ISAC. But there may be some omissions, largely because of fading memories or lack of my exposure to certain instances. I sincerely apologise for any such omission.

Bengaluru, India                                        Prem Shanker Goel

# Acknowledgements

The inspiration to write this book came from Dr. B. N. Suresh, a colleague and a friend who had completed his book on design of launch vehicles and wanted me to write a book related to the evolution of satellites in India. I thank him for his continued persistence. Writing this book on Making of the Satellite Centre gave me an opportunity to recall my association with Prof U. R. Rao and live those moments again. I am grateful to him for those wonderful memories. There are hundreds of persons who have made their contributions in this journey, some alive and many passed away; each deserves a big *thank you*, whether captured in the book or not.

Dr. Surendra Pal and Mr. V. R. Katti reviewed the manuscript and made numerous suggestions, particularly in recalling the names and specific contributions of some of the persons that I missed in my first draft. I thank both of my colleagues for their patience and contributions. Mr. Guruprasad helped me with the chronology of events and in capturing photos of important milestones.

Indian National Academy of Engineering (INAE) decided to make this book a part of their publication and it was a big morale booster. I thank the engineering academy for all the support.

Prem Shanker Goel

# About This Book

This book is all about the people who made ISRO Satellite Centre (ISAC), what is U R Rao Satellite Centre (URSC) today, as seen by one who lived through initial years of the "Centre in the making" and watched things happening, in addition to making his own humble contributions.

It was a dream of one man, shared by many and contributed by many more, Prof U R Rao will live as long as the Satellite Centre exists. The landmark journey from a small Division (Satellite Systems Division, SSD) of the then Space Science and Technology Centre, SSTC (now Vikram Sarabhai Space Centre, VSSC), with a staff of fewer than 10 in 1970 has grown to more than two hundred times as the second biggest Centre of ISRO. Professor Rao is to URSC what Prof Sarabhai is to ISRO. Professor Rao not only gave the vision but also inculcated the work culture of ISAC and built the organisation. His passion for space, concern for Indian science and love for ISAC are difficult to describe in words. It is remarkable that he transferred that spirit and commitment to dozens of youngsters, who helped in making the Centre and preserving those values for decades to come.

It is all about wonderful work of wonderful people, where passion drives and nation thrives.

# Contents

# About the Author

**Dr. Prem Shanker Goel** pursued his B.E. in Electrical Engineering from University of Jodhpur, M.E. in Applied Electronics and Servomechanism from Indian Institute of Science (IISc), Bangalore, and Ph.D. from Bangalore University, India. He developed spin axis orientation system Bhaskara I & II satellites, magnetic control for spinning Rohini series satellites, momentum biased 3-axis control system for APPLE, zero momentum biased 3-axis control system for IRS, and V configuration momentum biased attitude control system for highly stabilized INSAT-2. He developed very agile control system with step and stare capability for spot imaging mission TES and guided the evolution of reentry capability through SRE mission. He was President of Indian National Academy of Engineering (INAE) and Vice President, Aeronautical Society of India, Dr. Vikram Sarabhai Distinguished Professor at ISRO. Dr. Goel was awarded the prestigious Indian civilian Award Padma Shree in 2001. He received several other awards including Distinguished Scientist Award of ISRO and lifetime achievement award of INAE. He is Fellow of Indian Academy of Sciences, Bangalore; National Academy of Sciences, Allahabad; Indian National Science Academy (INSA), New Delhi, among others. Currently, Dr. Goel is Honorary Distinguished Professor at ISRO HQ, Chairman, Comprehensive Technical Review Committee, GEOSAT Program, and Chairing Apex Committee for planning Communication, Remote Sensing and Navigation satellites of ISRO. Dr. Goel has contributed significantly to mission planning for remote sensing, communication, and scientific missions and authored over hundred research papers in referred journals and conferences.

# Chapter 1
# The Early Days

On 17 August 1970, I arrived at Thiruvananthapuram (then Trivandrum) railway station and hired an auto straight to Space Science and Technology Centre (SSTC), ISRO, with baggage, from the comfortable campus of the Indian Institute of Science, Bangalore, after completing my M.E. formalities just the previous day. The journey to the city was more comfortable as a senior engineer of SSTC gave me a lift in his car. I was asked to join Satellite Systems Division (SSD) as control engineer and it was my first disappointment. We knew Thumba (TERLS, the Thumba Equatorial Rocket Launching Station) only for rockets and not working on rockets did not look that fanciful. The SSD, located in the main building, had about 10 engineers, reporting to Mr Tarsem Singh as acting Division Head. Our Division Head, Prof U R Rao, visited SSD only during weekends, twice a month. He was conducting his research in X-ray astronomy at PRL Ahmedabad. I was in turn asked to report to Mr R Ashiya, who informed me that I had to work on Attitude Control System of spinning satellite, 40 kg RS-1 (Rohini Satellite 1), to be launched by SLV-3. Further, I should go to a library and find out what to do?

TERLS was established in 1963 by Prof Vikram A. Sarabhai, for launching sounding rockets to conduct scientific studies in the upper atmosphere and Thumba (near Thiruvananthapuram) was chosen because it was a less-populated land along the coast and was close to the magnetic equator. He took over as Chairman Atomic Energy Commission after the sudden death of Prof Homi Bhabha in an air accident in 1966. With his vision for Space, Prof Sarabhai created the SSTC/ISRO in 1967 and divided the SSTC into functional divisions to make a Satellite Launch Vehicle and also a division for making Satellites, the Satellite Systems Division (SSD). The team at SSTC had earlier studied a few configurations of the Satellite Launch Vehicle and the configuration No.3 with four solid stages was chosen for taking up the development, and hence, the name SLV-3. Professor Sarabhai's visits as Director were short and infrequent as he was heading some 28 institutions along with PRL (as Director). The SSTC was managed by the Technical Coordination and Finance Committee (TCFC) comprising five senior scientists with its head by rotation as Acting Director, in the absence of Prof Sarabhai. In those days of Telex

© Indian National Academy of Engineering 2022
P. S. Goel, *Making of a Satellite Centre*,
https://doi.org/10.1007/978-981-16-3480-2_1

era when even fax was not available and telephone connectivity to different cities could take hours and hours, it was amazing that Prof Sarabhai was there in every decision-making, in all of his 28 institutions, and the system was working very well. Dr M K Mukharjee, a material scientist, Dr V R Gowarikar, a nuclear and chemical engineer working on the development of propellants, Dr S C Gupta, a control engineer heading navigation, guidance and control, Dr A E Muthunayagam, a propulsion engineer and Dr Y J Rao an aeronautical engineer were the five TCFC members. In addition, Mr M R Kurup was heading Rocket Propellant Plant (RPP) and Mr H G S Murthy was the Director of TERLS, responsible for the launching of sounding rockets.

The SSD had a bumpy start with Mr B Rama Krishna Rao as its Head in 1967. He was also heading Development of Electronic Systems (ELDR) and Electronics Production (ELP) divisions for the launch vehicles. I understand that there was a protest from a few engineers in the small SSD, to Prof Sarabhai and subsequently Mr P P Kale, another scientist at Physical Research Laboratory (PRL) Ahmedabad was appointed as its head. An opportunity came to have a joint study for a communication Satellite for India (INSAT-1) with Massachusetts Institute of Technology, USA, and Mr Kale left for the US to lead the ISRO-MIT study. Mr R M Vasagam, Mr Y S Rajan and Mr K Narayanan were also in the study team.

Professor U R Rao, then professor at PRL, was made the Division Head of SSD in 1969, and this perhaps was the beginning of the ISRO Satellite Centre in the making. There were seven engineers in SSD in 1970 before I joined. Mr Tarsem Singh, acting head, was also heading satellite Instrumentation, Ground Check Out and Integration. Mr M K Saha was working on RF systems (Transmitter and receiver), Mr R Ashiya on Telecommand and encoder, and Mr D V Raju and Mr B B Verma on Telemetry. Mr P Radhakrishnan was working on Power Systems. Mr Ravindranath Raut was assisting Mr Saha on RF systems. Then, Mr R Sellappan, Mr K Thyagarajan, Mr D Venkataramana, Mr O P Sapra and myself joined together within a span of a few days of each other in August bringing the SSD strength to 12. Mr Thyagarajan joined Mr Raju for development of Telemetry and Mr Venkataramana as digital engineer for Telecommand. Mr Sellappan and I started as Attitude Control System engineers. Mr Sapra joined Mr Tarsem Singh for Ground Check Out.

SSD's immediate mandate was to develop a 40 kg satellite, RS-1, to be launched by SLV-3, into a 400 km, circular, 44 degree inclined orbit from Shriharikota (SHAR) launch range, which was under development. Dr Y J Rao was Project Engineer for development of launch Complex at SHAR. RS-1 was a technology mission and had to be a simple, spinning satellite, with all basic systems and to be operational for one year. 1970 was the year of the first transformation as the strength of SSD was more than doubled in a span of 6 months. New expertise/disciplines were being added every week. To start with, there were only communication (RF), Telemetry/Telecommand, Power and instrumentation disciplines. Then came attitude and orbit control and thermal control systems. Later in 1970, Mr R S Mathur and Mr S Y Ramakrishna joined Mr Radhakrishnan for solar panels and power systems, respectively, Mr V K Kaila, Mr V D Prasad for Thermal system, and Ms Rajalakshmi joined Mr Raju, and Mr V Gopala Rao joined Mr Saha.

In 1971, Spacecraft integration and Structure were added. Mr V R Pratap joined Mr Sapra as Ground Checkout engineer. Mr S Pal and Mr Kista Reddy as antenna engineers, Mr T K Alex for Attitude sensors and instrumentation, and Mr V K Kaila for thermal control. Later, Mr J P Gupta joined Telecommand. Mr V A Thomas, Mr A V Patki, Mr H Narayanamurthy and Mr Srinivasan were transferred from different divisions of SSTC to SSD. Mr Thomas was made head mechanical facilities, Mr Patki, Head, Structures. Dr Subramanyam from SSTC joined Structures group and Mr Narayanamurthy as head Thermal systems. Mr P S N Rao and Mr C Kameshwara Rao joined control system to work on control electronics. Mr S Kalyanaraman, Mr M L N Sastry and Mr R N Tyagi joined RF Systems. Mr V Gopala Rao initiated to develop Transmitters. Mr R Seshaiah and Mr S M Bedekar joined Telemetry group. By the end of 1971, we were a compact Satellite Systems Division of about 35 engineers, representing every discipline of the satellite technology.

Professor U R Rao was looking beyond RS-1, to do more from satellites, like remote sensing and astronomy. Always thinking of satellites and payloads, one Sunday morning, he drove to our house (we four bachelors from SSTC were living together in a rented house in the city) just to discuss precise control of a gas in an X-ray counter and how to sustain it in the orbit for a long duration. Sometime in 1971, during a chat about the future of SSD, he said that very soon we would have over a hundred engineers working on different aspects of satellite technology in SSD. The seed to form a satellite centre was there (in his mind) and it was awaiting an opportunity to be sown.

The year 1971 ended with a tragedy. Professor Sarabhai expired on 30 December 1971, due to a heart attack, in the night at his hotel at Kovalam beach, Trivandrum, after a full day of hectic activity. Just in the evening, he had delivered a speech to SSTC employees on the open ground floor of the main SSTC building (the building stands even today, with its old charm and glamour). It was the first and last opportunity for me to listen to Prof Sarabhai in a general address, though I had met him earlier when Prof Rao brought him to show SSD. When Prof Rao introduced me to Prof Sarabhai as a young control engineer, Prof Sarabhai instantaneously said, "Keep good relations with Dr Gupta's division". I could not understand the meaning of that advice at that time but the words remained in my mind. It was found to be very useful later on for decades during our interaction between SSD and Control, Guidance and Instrumentation division, and Dr Gupta in particular.

I remember only one statement in his speech as it related to us. He told the management, "Take care of young engineers at 'C' and 'D' level, and they will take care of the work". The sudden demise of Prof Sarabhai was a huge setback and everyone was worried about the future of ISRO. Professor M G K Menon was appointed as Chairman ISRO with the understanding that it was only a temporary arrangement. However, what followed soon after was yet another event that changed the destiny of SSD and ISRO.

The President of the USSR visited India in February 1972 and offered Prime Minister, Mrs Indira Gandhi, to help India in building an Indian Satellite and also to be launched by a Soviet launch vehicle, as a friendly gesture. It was perhaps just to strengthen ties between our two countries in those times of the cold war in

**Fig. 1.1** Lab pooja at Peenya

the bipolar world. Mrs Gandhi was a leader; she did not take even a second to say "YES". Professor Rao was called by the evening over the phone to ask how much it will cost to make a Satellite and he instantaneously said rupees three crores. Project ISSP (Indo Soviet Satellite Project) was thus born with this one call. Professor Rao took a schedule of 3 years on himself and decided that the project ISSP should move to Bangalore for speedy development, keeping in mind the industry and academia support at Bangalore. As planning and understanding the working interface with the Soviet Union was under way, Prof Satish Dhawan took over as Chairman ISRO in June 1972, on returning from CALTECH (California Technical University, USA), after completing his assignment as Visiting Professor. He too approved of the movement of the ISSP to Bangalore. There was some political opposition for shifting ISSP from Thiruvananthapuram, but Prof Rao, with the support from Prof Dhawan and Delhi, managed without much difficulty. We all now know how correct that decision was!

Even before ISSP, Prof Rao had been strengthening SSD with the induction of Engineers at all levels. Dr S P Kosta joined as Antenna expert from CEERI Pilani, becoming number two to him. He had also appointed a few engineers at PRL, Ahmedabad. Mr K S V Seshadri, Mr B L Agrawal and Mr A D Dharma were working in his Cosmic Ray lab. Mr V R Katti, Y K Jain, Mr Arun Batra, Mr V Jayaraman, Mr Srinivasa Murthy, Mr Ganage, Mr S Dasgupta and Dr Gambhir were working in SSD/SSTC, PRL. With the approval of the state government of Karnataka, he got allotted six Industrial sheds (A1–A6) which were under construction in the emerging industrial area at Peenya to ISSP. A lab pooja was conducted on Ganesh Chaturthi day, on 11 September 1972 (Fig. 1.1). This day is considered as the day of birth of ISAC (ISRO Satellite Centre) as it marks the beginning of a vision that goes much beyond making payloads for Launch Vehicles.

# Chapter 2
# The ISSP

Setting up of Indo Soviet Satellite Project (ISSP) in the under development Peeniya Industrial Estate at Bangalore coincided with the setting up of ISRO Head Quarters (HQ) in the old Gymkhana building of Indian Institute of Science (IISc) as Prof Dhawan had decided to make Bangalore as ISRO Head Quarters and simultaneously retaining his position as Director, IISc (till he retired from the directorship of IISc in 1982). Professor Dhawan decided to create Department of Space (DoS) as an administrative body on the lines of Department of Atomic Energy and Space Commission like Atomic Energy Commission, an empowering body. Mr M A Vellodi joined as additional Secretary and Mr T N Seshan joined as first Joint Secretary in the Department of Space to organise the administration. Professor Brahm Prakash was appointed as Director of SSTC, Trivandrum. It was renamed as Vikram Sarabhai Space Centre (VSSC) in memory of the founder. An IAS officer and Ex controller of BARC, Mr Kulkarni was appointed as Controller of VSSC as head of administration.

A new Centre, the Space Application Centre (SAC) was formed in Ahmedabad, merging ESD, SCSD, MID and ESCES divisions with Prof Yash Pal as its Director. Restructuring of ISRO as a formal organisation, bringing it under the Department of Space (DoS), Government of India, and later (in 1975) converting the whole of ISRO into a government body from the status of autonomous society were very fundamental changes brought about by Prof Dhawan. ISRO under Prof Sarabhai was more of an informal autonomous body running under directions from Prof Sarabhai and administered by Mr Thakur at PRL, a confidant of Prof Sarabhai. It was indeed a far-sighted step to start with, but as a growing organisation with its budget growing at an exponential rate, Prof Dhawan realised that it had to be converted into a government body. DoS and ISRO HQ were shifted to Cauvery Bhavan, a new building at Mysore Bank circle on Kempegowda Road, Bangalore. What we see today as ISRO is the vision of Prof Sarabhai, but moulded into a vibrant system by Prof Dhawan.

The first seminar on ISSP was organised in Bangalore in July 1972 and the team SSD travelled to Bangalore to make its first-ever presentation to Chairman ISRO, Prof Dhawan. It was a bad start as the luxury bus in which we, about 30 engineers, were travelling to Bangalore toppled near Alwaye (Kerala). Mr R S Mathur broke

© Indian National Academy of Engineering 2022
P. S. Goel, *Making of a Satellite Centre*,
https://doi.org/10.1007/978-981-16-3480-2_2

**Fig. 2.1**   1972 photo of Peenya sheds

his both collar bones. Luckily, there was no other major injury for any of the other team members.

In spite of the accident, we made it to the seminar and Prof Dhawan was happy listening to the team's presentations for the whole day. The project ISSP had taken off. Soon, the first batch of engineers from SSD moved to Bangalore during October–November, 1972. Most of the engineers were shifted from SSD to ISSP and over the next few months, new engineers were recruited in large numbers. Mr T C S Chaudhary was appointed as administrative officer. However, symbolically, a small team of SSD was retained at SSTC, Trivandrum, with Mr Tarsem Singh as in charge. Mr Tarsem Singh, Mr T K Alex and Mr S M Bedekar also later moved to ISSP after a few months but Mr Radhakrishnan remained there as head SSD for a few years. Professor Rao decided to shift himself to Bangalore and devote full time to ISSP. Also, both teams at PRL, the Cosmic Ray lab as well as SSD/SST, PRL, were shifted to ISSP Bangalore.

We first moved to Peenya industrial sheds A1–A6 (Fig. 2.1) and it was decided that A5 and A6 will primarily be an electronics laboratory. As a part of electronics lab, a special enclave for a cleanroom was created for satellite integration. A one-metre thermo-vacuum chamber was installed in the electronics lab. Dr S P Kosta was designated as Principal Technical Officer.

To build a satellite laboratory out of empty, unfinished industrial sheds was challenging and exciting. Professor Rao caught hold of one consultant, Mr P N V Rao, who was the answer to all the problems: civil, electrical, plumbing, whatever. While effort was on to setting up laboratories and test facilities, periodic meetings with Soviet experts through INTERCOSMOS began. Professor Koftunienko was appointed as Project Director ISSP from Soviet side, a counterpart to Prof U R Rao. Soon, Dr K Kasturirangan also moved to Bangalore from PRL, to take charge of coordinating scientific payloads for ISSP and was additionally appointed as member secretary of ISSP Project Management Board. It was chaired by Prof Dhawan. Professor Brahm Prakash was another key member. At that time, an administrative question was raised regarding the status of ISSP. The project ISSP belonged to which institution? Senior management at VSSC wanted it to be a project of VSSC but Prof UR Rao did not agree. He was convinced that ISSP must have its separate identity, as per his larger dream of a separate satellite centre. However, he agreed to report to Prof Brahm

Prakash in his individual capacity, but not as Director, VSSC. In practice he was only reporting to Prof Dhawan. This was a major step towards creating an identity for ISSP.

We had many new important manpower additions in the beginning of ISSP. Mr N K Malik joined me as control engineer and Mr Sathe as mechanical engineer in Control System group. Mr M G Chandrasekhar joined program planning. Mr Y K Singhal and Mr R D Bansia joined Telecommand; Mr Lakshmisha, Mr K N S Rao, Mr V Mahadevan, Mr K N Shamanna, Mr Parthasarathi and Mr Manthaiah joined Communication group; Mr D R Bhandari, Mr P P Gupta and Mr Bhojraj joined Thermal, Mr T K Goswami Mechanical Integration and Mr A A Bokil Electrical Integration. Dr P S Nair and Dr M S S Prabhu joined structures group. Mr S Dasgupta took responsibility of getting structure manufactured at HAL. Dr A S Prakasa Rao, the first student of Prof Rao, was also transferred from PRL a little later to take care of mission aspects. Mr R S Bhat joined to work on orbital mechanics and orbit determination and P Padmanabhan transferred from PRL to look after attitude determination. Mr P J Bhat joined to manage satellite data processing. Dr N Ramani, a student of Prof Dhawan, joined as computer expert to manage computer facility and provide software support. With him, Dr R Sundar organised the computer facility and software support to users like structures and thermal groups. Mr Eshwara Prakash joined Mr Sethunathan to develop mission software for satellite management, thus covering all aspects of mission. Dr R D Gambhir transferred from PRL was assigned to organise Environmental Facilities. Mr B N Baliga joined Dr Gambhir soon after. Dr Pathak, last student of Prof Sarabhai, who later completed his PhD with Prof Rao, came as Head Program Planning and Evaluation Group. Dr D P Sharma, another student of Prof Rao at PRL, came to assist Dr Kasturirangan on space physics in general and to work on X-ray astronomy payload of Aryabhata in particular. Dr T M K Marar joined as head Quality Assurance Group. Professor Rao created formal organisation with group heads reporting to him.

The design and development of satellite systems started and schedules for the project were drawn. Most of the electronics development that took place for RS-1 was adopted for ISSP. The Ni-Cad battery, solar panels onboard tape recorders were to be supplied by the Soviet side and the power electronics to be developed at ISSP. The Telecommand and telemetry systems, VHF transmitter and VHF receiver, instrumentation like sun sensors and magnetometer were being developed at ISSP. Mr K Thyagarajan suggested the use of CMOS devices, developed by RCA for their low power consumption and better noise immunity. ISRO was perhaps the first organisation in the world and ISSP was the first satellite to use CMOS technology.

The structural engineering model was delivered from Hindustan Aeronautics Limited (HAL) during a meeting with the Soviet team. Dr Uarof, an expert from USSR, shook it by his hands and said, Nyet (No in Russian). He then sat overnight to produce a design of bottom shell of the satellite structure interfacing with the launch vehicle which was subsequently fabricated at HAL. Cold nitrogen stored gas spin-up system was also supplied by Soviets. Our role was primarily on the analysis of spin control and testing. We were to test pyro cartridges through a test console supplied by the Soviets. I wrote a one-page procedure and the Soviet team said "Nyet" again and wrote a 5–6 page note along with a log book. In the USSR, testing operations were conducted by young army men, who could only read and write but have no idea what it means. Even turning a knob to ON/OFF position has to be explicit to the extent of turning the knob right or left, 45 or 30° and a bulb glowing green or red. Each and every operation, however small, had to be documented, signed and counter-signed by Quality Assurance (QA). This perhaps was the most important lesson in Space Technology. We designed and fabricated a fluid-in-tube nutation damper, partially filled with mercury in a stainless steel toroidal tube. By the end of 1973, infrastructure was mostly in place except for the one-meter thermo-vacuum chamber, and we were busy in translating bread board designs into flight worthy packages. Professor Rao realised that the six sheds from A1 to A6 were not adequate and additionally, B-type sheds opposite A1 and A2 were acquired. A common electronics laboratory was established in sheds A5 and A6. Each subsystem was allotted two tables with a separate group for wiring the circuits. This lab was the centre of activity and interest. Chief Controllerate of Inspection (CIL), a unit of defence, was utilised to screen electronic components. M/s Hegde and Gole was developed to fabricate Printed Circuit Boards for the project. Environmental test facilities like vibration shaker as well as hot and cold chambers were installed. A modern but small mechanical fabrication facility (Workshop) was set up in B-type sheds. Besides, an anodising facility and a thermal paint booth were set up at HAL for thermal work.

It was also time to think ahead. LANDSAT 1 and 2 satellites of NASA were making waves in remote sensing and a question, can we quickly design a remote sensing satellite? was raised. Challenge was posed to build a remote sensing satellite with as much use of existing ISSP technology. A study committee was constituted to arrive at a possible way of converting the ISSP into a remote sensing satellite in a minimum time. It was an opportunity for me to interact with possible remote sensing payload configurations and to think of satellites as a whole. Dr George Joseph at SAC came up with a configuration of a payload using a Vidicon tube, registering the image like a photograph with a shutter opening for 2 ms. We configured a remote sensing satellite, on a spinning platform, not done hitherto by anyone else and providing imagery in two bands. Mr O P N Calla, from SAC, Ahmedabad, came up with the proposal of passive microwave radiometers (3 bands) to further augment the remote sensing applications.

Meanwhile, the work on ISSP was going on at a very fast pace; most engineers were working till late in the evening and having periodic meetings with the Soviet side by rotation in Bangalore and Moscow. Mr Ashiya and Mr Venkataramana designed the Telecommand (TC) system. Mr Raju and Ms Rajalakshmi designed the telemetry (TC) system. Mr Saha and Mr Gopal Rao designed RF (Radio Frequency) TM receiver and Mr Bhaskaranarayana along with Mr U N Das and Mr N N Singh designed the ground TC transmitter, both in VHF band. Dr Surendra Pal and Dr Kosta designed the on-board antenna system. Mr S Y Ramakrishna and Mr Radhakrishnan designed the power system electronics. Mr T K Alex designed sun sensors and magnetic sensors. Mr V A Thomas took charge of overall satellite integration while M N Sathyanarayana and Mr V R Katti took care of mechanical and electrical integration, respectively. Mr A D Dharma, Mr O P Sapra and Mr V R Pratap developed a Ground Checkout system under Mr Tarsem Singh. Mr A V Patki was assisted by Dr P S Nair and Dr M S S Prabhu for structure design. Mission responsibility was with Dr Prakasa Rao with Mr R S Bhat for orbital mechanics, Mr P Padmanabhan for attitude determination and Dr N Ramani and Mr P Sethuraman for data processing. The three scientific payloads were, i) Gamma ray payload from Tata Institute of Fundamental Research (TIFR), with Principal Investigator (PI), Prof Daniel and Prof Damle, ii) Aeronomy payload from PRL with PI Prof Satya Prakash and Dr Subbaraya and iii) X-ray payload from ISSP its PIs being Prof U R Rao and Dr K Kasturirangan. All the payloads were coordinated by Dr K Kasturirangan with Mr V Jayaraman providing assistance. Mr K Thyagarajan was coordinating components and other interface issues. As these payloads used high voltage, hence, arching at critical pressure was an important concern to be watched and resolved. Facilities for assembly, integration, testing and qualification were set up as the satellite ISSP was getting ready. The name of the project underwent change two times, retaining ISSP in short form: (i) Indo-Soviet Satellite Project, (ii) ISRO Scientific Satellite Project and finally (iii) Indian Scientific Satellite Project.

The project was to make use of ISTRAC (ISRO Satellite Tracking Unit) Ground Station at SHAR (Shrihari Kota launch complex, 100 km north of Chennai). We also set up our own equipment at the Soviet ground station, Bears Lake, near Moscow, to provide additional support during the mission. This station support continued even for subsequent Bhaskara and IRS missions.

While the satellite was in a thermo-vacuum chamber, Dr Kosta proposed to make a backup ground station with a manually operated Yagi antenna. As no place was available in those sheds, a lavatory was quickly converted into a Ground Station Laboratory by Dr S Pal and Dr Lakshmisha. Needless to say that this backup ground station served very well for a long time till the VHF system continued to be in use in ISRO satellites like Bhaskara and APPLE.

The ISSP satellite was about to be shipped and Prof Dhawan proposed to give it a name. Three names (Jawahar, Netra and Aryabhata) were suggested to the Prime minister Mrs Gandhi and she chose to name it as Aryabhata, after its launch, a practice many countries follow to give a name to a satellite after the launch.

INTERCOSMOS was a big, magnificent vehicle and the launch operations had gone very smoothly, particularly for a first satellite. The launch site in the Soviet Union, Kapistuniya, was quite cold in March/April, but the warmth of sunlit weather on 19 April 1975 was to write a new page of history for India. The satellite was launched southward and separation was confirmed. The satellite was powered off during launch and was to be automatically powered on following separation from the launch vehicle. The ISTRAC station at SHAR received clear signals. Aryabhata (Fig. 2.2) was born and arrived. India had joined the space club.

Aryabhata became an inspiration to the country and the only comparable event was the Pokaran-I, in the previous year, 1974, another brainchild of Prof Sarabhai. Professor Sarabhai had targeted the launch of SLV-3 also in 1974 and his planning 5 years ahead could not just have been a coincidence. He was the only common link between the two agencies, Department of Atomic Energy and ISRO.

**Fig. 2.2** PM visiting
Aryabhata Satellite in
cleanroom

**Fig. 2.2** PM visiting Aryabhata Satellite in cleanroom

However, there was a setback; the satellite did not spin after ejection and the power supply to the three payloads failed soon after the payloads were switched on, hence, no meaningful scientific data could be collected. The satellite, though functioned very well from a technology point of view, was a great success, as all the mainframe systems worked well and were proven in space till it remained in orbit for about 17 years.

The story behind Aryabhata is not just the development of technology or a piece of hardware. It is the story of a set of people who committed themselves to do something that looked impossible. Aryabhata was built in two and a half years after moving to Bangalore in unfinished industrial sheds. Only a very few were above 30 years, the average age being 27 years, and none other than Prof Rao, who had put his X-ray astronomy payload in the Pioneer series of spacecraft while doing his postdoctoral research in Texas, US, had ever seen a satellite. Even the Soviets never agreed to show us any satellite or even a test facility even though we were meeting often, at least once in a year in the Soviet Union. All meetings were held in INTERCOSMOS office in Moscow. This team of about 200 engineers and technical staff was experimenting, debating and innovating every day. A new culture was taking birth, in which a young engineer/scientist could challenge a senior on technical issues, without carrying a feeling of fear or disrespect.

The administration was as committed as engineers, to do their bit in providing facilities or taking action for procurement. Internal Finance Advisor (IFA) was part of the team and advising on how to quickly organise the lab within the framework of rules. It is this harmony and the commitment that had never been seen in the country and could never be repeated even in ISRO, except, perhaps, in Technology Experimental Satellite (TES), some 25 years later. Aryabhata was a simple satellite from today's standards but, it gave India confidence that sophisticated technology could be done in the country. What followed later in sectors like IT or automobiles owes a lot to that confidence created by Aryabhata.

# Chapter 3
# The Experimental Era of Spaced-Based Services

## 3.1 Beginning of Remote Sensing

ISSP team was keen to convert Aryabhata to Satellite for Earth Observation (SEO) and the project report was quickly prepared. It was a big jump in technology as not only the state-of-the-art payloads had to be built, but also the satellite had to be significantly modified for pointing the payload towards earth for collecting imagery and transmit the payload data to the ground stations, needing higher data rate and higher power of the transmitter.

To point the payload towards the earth, it was planned to orient the spin axis of the satellite perpendicular to the orbital plane and spin the satellite from 5 to 11 rpm. The camera was mounted on the satellite in the belly, perpendicular to the spin axis, hence the camera would look at the nadir once in every spin cycle. The two-channel camera shutter would open for 2 ms while looking vertically down. The microwave radiometers were mounted on the opposite side of the camera, thus looking at the earth half-spin cycle later. The spin rate was so adjusted to give sufficient overlap of the two successive images.

The cold gas system of Aryabhata was modified incorporating a pressure regulator, spin axis control thrusters (with solenoid valves) and a unique 8 pulse logic was designed to precess the spin axis with practically no nutation component leftover. This was achieved by adjusting the moment of inertia of the spacecraft such that the transverse velocity given by the first pulse is in phase opposition and cancelled by that of the fifth pulse, leading to pure precession after 8 pulses. Automatic control logic was incorporated by computing the difference of the two horizon sensor pulses, placed 45° from the Nadir. SEO was perhaps the only satellite in the world with closed-loop control of the spin axis in a spinning satellite. A fluid-in-tube nutation damper was also developed filling silicone oil into a fibre-glass toroidal tube. A magnetic bias control was introduced for extending the mission life of the satellite by compensating the 4.6 deg. Orbit plane change.

There were many new technology developments like new antenna configuration, and of course the qualification of the payload under possibilities of corona as the

© Indian National Academy of Engineering 2022
P. S. Goel, *Making of a Satellite Centre*,
https://doi.org/10.1007/978-981-16-3480-2_3

payload had high voltage supply for the camera and high bit rate telemetry for transmitting payload data.

SEO was a forerunner for ISRO to get into an era of remote sensing. Dr Kasturirangan was appointed as Project Director, and Dr Kamath (SAC) took the responsibility of data utilisation and data products. The ground facilities and methodologies of utilisation were evolved using data from LANDSAT 1 and 2 satellites and were applied to problems like detection of wilt disease in the coconut trees, right in the beginning stage, before it seriously affected the trees. National Remote Sensing Agency (NRSA) at Hyderabad was brought under DoS and it oriented itself towards bringing these capabilities of using our own satellite data and marketing it.

The young Aryabhata team was getting mature and new engineers were joining continuously. Mr Saha organised the RF team with Mr V Gopala Rao, Mr S Kalyanaraman, Mr M L N Sastry and Mr Sivaprasad. Dr Surendra Pal with Mr Kista Reddy, Mr Lakshmisha, K N S Rao and Mr Mahadevan formed an antenna expert team. Mr R Ashiya was in charge of Telecommand, with him Mr D Venkataramana, Mr J P Gupta, Mr Y K Singhal and Mr U N Das as key digital designers and Mr Bhaskaranarayana developing TC transmitter. Mr Thomas was Head integration with Mr M N Sathyanarayana and Mr V R Katti as his deputies. Mr B L Agrawal was responsible for power electronics and Mr R S Mathur for solar panels. Mr M G Chandrasekhar returned to PPEG tasks from electronic fabrication and mission assignments of Aryabhata.

A space grade Central Electronics Fabrication Facility (CEFF) was formed with Mr V R Katti as in charge with Mr R Chandrasekhar and Mr Gopalkrishnan supporting him. Mr Siharan De emerged as an expert for space grade components and Mr K R Ramagopal took charge of Quality Assurance as Dr Marar moved to Technical Physics as its head. Mr S Nagabhushnam took charge of the development of the cold gas propulsion system and Mr S Murugesan as designer of control electronics. Dr Prakasa Rao along with Dr V S Iyengar, Mr Y N Bhushan and Mr S K Shivakumar looked after the mission. Professor Rao was completing the jigsaw puzzle, putting every piece at the right place; a complete satellite centre was in the making through SEO.

In addition to the design team, it was important to build facilities for Spacecraft subsystems and systems testing, fabrication facilities for mechanical and electronic hardware, components screening, screening of Printed Circuit Boards and all the test and evaluation capabilities at component, subsystem and the system levels. Professor Rao was clear that these projects were the fertile grounds for building up the capabilities and expertise that one day would bring India into the forefront of Satellite Technology. However, with all these efforts, the dream of creating a satellite centre still remained elusive.

The Payload development was taking place at SAC. Dr George Joseph was the Project Manager (PM) for the camera payload with Mr A S Kiran Kumar, Mr Kaduskar, Mr Nagarani and Dr Ram Ratan as key engineers, with Mr O P N Calla as PM for Microwave Radiometers with Mr S S Rana, Mr S Balasubramanyam and Mr G Raju supporting the development.

The SEO (Fig. 3.1) integration and testing was smooth except for noise problem in

**Fig. 3.1** SEO, Bhaskara-1
in cleanroom

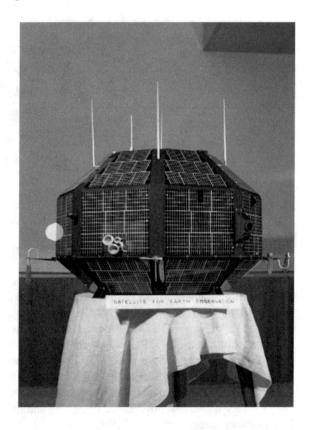

the payload data. It was a typical EMI problem threatening the launch postponement. However, the team under Mr V R Katti could locate the coupling lines and solved the problem through isolation, shielding and use of power line filters. This later led to generating new integration guidelines.

The SEO-1 was launched on 7 June 1979 and given the name Bhaskara-1 after ancient mathematician and astronomer Bhaskaracharya I. The satellite worked well, but the Videocon camera got switched OFF soon after turning ON, all by itself, creating some disturbances in the satellite. After all the analysis, it was concluded to be a corona discharge in the high-voltage power supply as the cause. As the power supply was potted, obviously, it had trapped air bubbles with a leakage path and there was no way to know how good or bad the leakage path of the trapped air was. A spare package was subjected to extensive thermo-vacuum testing and the problem was simulated on ground. It was decided to wait for almost 8–9 months assuming a very low leakage rate and the camera was switched ON again. The real-time display came alive giving very good-quality pictures of the Indian landscape. A new era began, the Experimental Remote Sensing era! Bhaskara picture resolution was coarse, about one kilometre, but the complete methodology of remote sensing

**Fig. 3.2** An invaluable award

from data collection, data processing and data archival to information retrieval was being established, hence called the experimental era.

I have a personal sentiment attached to this event. On that day of switching ON the payload, Prof Dhawan saw me in the morning and asked what would happen? I said it will work. "Shall we have a bet?" he asked. I was a bit taken by surprise but said, "yes sir". He said OK, five rupees. I said fine sir but five rupees is too small an amount for Chairman ISRO to bet. He said, "What do you mean? It is half a year's salary your department gives me". He was drawing only one rupee as token salary from DoS/ISRO as he was drawing his salary from IISc as its Director. Well, I demanded my five rupees soon after the pass and requested him to sign on it. This is one of the most precious possessions of my life and I am preserving this five rupee note as a very special gift of God (Fig. 3.2).

Bhaskara-1 served much beyond its intended life and Bhaskara-2, a repeat of Bhaskara-1, was launched on 20 November 1981 from the same Russian INTER-COSMOS vehicle, from Kapustinyar, near Volvograd. Both Bhaskara-1 and 2 gave the methodology and infrastructure for satellite remote sensing. To commemorate the occasion and to recognise it as a national achievement, Government of India printed the Bhaskara satellite picture on a two rupee note.

## 3.2  Experimenting with Satellite Communication

The seed of satellite communication was sown by the founder himself. A visionary and a charismatic leader summoning high esteem amongst the international community of space scientists, Prof Vikram Sarabhai had persuaded NASA (National Aeronautics and Space Agency, USA) to lend its most advanced satellite in the making, ATS-F, with a large unfurlable antenna, to India for experimental mass communication to the most underdeveloped regions. ATS-F was the most complex satellite having frequencies from VHF/UHF, S and C bands with multiple communication, broadcasting and MET payloads. It is amazing that Prof Sarabhai could convince

NASA to relocate this satellite over India even before the launch of the satellite. NASA kept its promise to the late leader and satellite ATS-F was brought to serve 8 clusters of the most underdeveloped regions of India under the program named as Satellite Instructional Television Experiment (SITE) during 1975–76. The idea was to connect these inaccessible underprivileged regions through satellite ATS-F and install community TV sets in the villages of the regions and transmit programs that are in local languages with regional content mixing education with entertainment (Fig. 3.3). A separate unit, Developmental Education and Communication Unit (DECU), was created at SAC Ahmedabad as there were no studios in the country which could produce such TV programs. Looking at the scale and involvement of a large population, SITE is considered to be the largest social experiment involving high technology, in the world. It is remarkable that NASA extended the services of one of its most advanced satellites, ATS-F, to stay at the Indian slot for one more year, perhaps the best tribute to Prof Sarabhai. SITE also provided an experience to ISRO on ground system operations and maintenance. DECU created the first TV program-making studio in the country. Every smart mobile phone is a TV studio today. But the spread of studios in the eighties resulting from commercial TV services because of INSAT-1B owes a lot to SITE and DECU. SITE teams installed and maintained thousands of receiving terminals in very remote places, travelling miles and miles on foot.

After ATS-F, at ISRO's request, a Franco-German Satellite "Symphony" was deployed over an Indian slot for continuing communication experiments in C band for another 2 years. A lot of experiments that included TDMA and CDMA techniques were conducted.

**Fig. 3.3** Concept of SITE

## 3.3   Exploiting an Opportunity—APPLE

The success of Aryabhata and the progress of SEO in the year 1975 instilled confidence in this young team to think big. The joint ISRO-MIT study team of INSAT-1 in
the US had completed its study in the form of a feasibility report for satellite configuration, mostly the payload, a mix of satellite communication, satellite broadcasting
and satellite meteorological observations through Very High-Resolution Radiometer
(VHRR) payload. Unexpectedly, we learnt that the European Space Agency (ESA)
had announced an opportunity for flying a satellite on its third developmental flight
(LO-3) of the upcoming launch vehicle Ariane-1, to be launched within 3 years. The
conditions were that the satellite had to carry a European Satellite METEOSAT on its
top and had to be delivered in time, if not, a dummy satellite to be built and delivered,
to carry the METEOSAT on its top. The opportunity was seized by quickly preparing
a proposal. Mr R M Vasagam, the then Head PPEG at VSSC, who in fact spotted this
opportunity and also prepared ISRO's response, was appointed as Project Director.
Mr P Ramachandran from VSSC was appointed as his deputy during the course of
the project run. The Project was named as Ariane Passenger Payload Experiment
(APPLE).

APPLE was configured with two C band transponders as a forerunner for ISRO
built INSAT's. A solid rocket motor, derived from the SLV-3 fourth stage was used as
Apogee Boost Motor (ABM) to push the satellite from a 7 deg inclined Geo Transfer
Orbit (GTO) to Near Synchronous orbit. A debate took place whether APPLE should
be a dual spinner or a 3-axis stabilised one. Based on the spinner experience of
Aryabhata and Bhaskaras perhaps, and because of Hughes Space Systems expertise
in dual spinners, Mr Vasagam was convinced that the dual spinner could only be done
in the time frame. However, the design team was of the view that operational INSAT
could not be a dual spinner for reasons of growth and flexibility, and hence, APPLE
should also be 3-axis stabilised. With a lot of discussions on enabling technologies
like the availability of monopropellant propulsion system and momentum wheel
v/s dual spinner antenna and the limited growth scope for dual spinners, 3-axis
stabilisation was accepted. It is important to note that Hughes Space Systems had
mastered the dual-spin stabilisation approach for INSELSAT class satellites as a
cost-effective alternative, though the rest of the agencies in the world had opted for
3-axis stabilisation. This one decision if gone otherwise would have put a lot of
doubts on our own capability to take on INSAT-2 series of satellites in the coming
years.

## 3.4   APPLE Enabled the Creation of ISAC

APPLE was a technology jump and a schedule challenge. Along with APPLE came
the recognition that ISSP had matured to take up a bigger role and it should no
longer continue as a project. Thus, the ISSP became ISRO Satellite Centre (ISAC)

on 3 November 1976. Professor U R Rao was appointed as its Director. This was a dream coming true and a joyous moment.

Professor U R Rao took command of APPLE in his hands to see that it was done in time. The TTC though remained in the VHF band; all the subsystems were newly designed for the Geo mission. The thermal control system design was much more involved for 3-axis stabilised satellite in a geostationary orbit and the structure was conceived as a thrust cylinder to carry heavier METEOSAT on its top. The monopro-pellant propulsion system was procured from M/S Hamilton Standard USA and the Momentum wheels were procured from M/S TELDIX, Germany. The AOCS team got busy designing the Momentum biased 3-axis stabilisation system with Magnetic control for momentum dumping and controlling roll/yaw error. Earth Sensor was procured from SAGEM, France, but there were no Gyros and it was decided to manage with pseudo rate damping using Pulse Width Pulse Frequency Modulator (PWPFM) controller. Mr Rajaram Nagappa was appointed as Project Engineer for the Apogee Boost Motor. Mr Leo Lasarado from SAC for payload, Mr N Vedachalam, Director IISU for Momentum Wheels and solar array drive, Dr K Anantha Ram, LPSC for Propulsion System, Mr VA Thomas for integration and Dr P S Nair for structure became part of the APPLE team. The VHF TTC system was developed by Mr Gopal Rao, Mr Shamanna and Mr Sivaprasad. The antenna was designed by Dr Pal's team consisting of Mr Lakshmisha and others. I was designated as Project Engineer, AOCS including inertial systems and propulsion. Dr P Kudwa joined as control engineer, to work on attitude control system as Mr Malik had been deputed to INSAT-1 project resident team. Weight control and schedule were very important issues and strict configuration control was organised to keep the weight under control and techniques like low-weight mechanical packages and high-density connectors were adopted. A weight optimisation committee chaired by Mr Ashiya, amongst other things, recommended Raychem wire in place of Teflon wire.

The second development flight of the Ariane, LO2, resulted in failure, and that gave us more time for the third flight of LO3 and APPLE. This period was very hectic with a busy schedule of reviews with European Space Agency (ESA). The satellite had to be qualified for compatibility to both, the Launch Vehicle at the bottom as well as to METEOSAT on top.

The Ariane LO3 launch on 19 June 1981 from French Guyana was a perfect one and APPLE was released into a nominal GTO orbit. We all assembled at SHAR for post-launch operations and our first job was to collect tracking data and predict the orbit. Mr R S Bhat from the mission team not only had put together the plan but also wanted to get confirmation from ESA experts, just to be sure. Plan to fire the ABM on the third Apogee was worked out and target attitude was determined. There were no gyros in APPLE, and the solid Apogee Boost Motor (ABM) was too big a rocket for 3-axis stabilisation, hence, spin stabilisation for ABM firing was the only option. Thus, APPLE had both provisions of spin and 3-axis stabilisation. APPLE was spun and oriented in the desired direction using sun sensors and magnetometers. The ABM performance was normal and next task was to de-spin, deploy solar panels and achieve 3-axis stabilisation, before arresting the spacecraft at the designated geostationary slot.

Of the two solar panels, one deployed nominally but the other did not and it was confirmed by thermal signatures as well. All efforts to deploy the second panel failed. We decided to go ahead with one panel not deployed for the 3-axis stabilisation, Sun and Earth Acquisition, the next day morning at 6 O'clock (Satellite Local Time). We had realised that the sun acquisition would not be easy as the one set of 4 pie sun sensors was covered by the non-deployed solar panel. Mr Y.K Jain and Mr Babukala Krishnan from the sensors team were busy in determining the attitude from the very slow, 256 bit per second VHF telemetry that we had in APPLE. We started very slowly, at the rate of less than 0.5 rpm (revolutions per minute), with spin axis perpendicular to the sun, but the satellite walked away without capturing the sun. The second attempt was also unsuccessful. We could not achieve sun acquisition and we had lost the opportunity that day. With a lot of overnight homework, we decided to reduce the spin rate even below to about 0.2 rpm next day, still no success. Had we lost the opportunity to 3-axis stabilise APPLE that day? It was as good as the mission lost.

Mr Babukalakrishnan made an observation, "Goel Ji, the satellite has walked away with more spin rate than it entered the sun field of view". That was a flash and gave a vivid picture to what had happened. We immediately organised a mini meeting requesting Prof Dhawan and Prof Rao to listen to the logic through phase plane diagram and assured that we will succeed in the next attempt. We would start with reverse spin rate and that would work. I am not sure even today whether they were convinced or not but both had confidence in my confidence or perhaps there was no other option, they gave "go ahead". The time window was running out, we reversed the spin rate, and slowly, at < 0.2 rpm, tried again. We had the sun acquisition. There was a big thunderous joy but we did not have the time to rejoice as we had to acquire the earth within the next few minutes (before 8.00 AM local time). We acquired the earth and spun the momentum wheel, transferred the control to momentum wheel and then started the long process of reducing the yaw/roll error by manual corrections. APPLE was now 3-axis stabilised, ready for orbit manoeuvres, for its location in Geo in the next 10 days. We had the C band transponder ON and for the first time, we were getting communication transponder signal at SAC (Ahmedabad) Earth Station from an Indian-built Communication Satellite (Fig. 3.4). The lavatory converted VHF ground station of Bhaskara was also updated as a back-up station to ISTRAC. A new era had begun. APPLE met all expectations in spite of one panel not being deployed and served for more than 2 years, until its fuel got exhausted.

APPLE is a milestone in ISAC's history and technology leap forward. 3-axis stabilisation using momentum wheels, static Earth Sensor, solar array drive, thermal control system of 3-axis stabilised satellite, solid Apogee Boost Motor, monopropellant propulsion system, qualification and testing of Geo satellites, Geo mission management and many other Geo-related aspects were put together. Though an experimental mission, it gave us all the necessary technology inputs to prepare for an operational mission. Satellite Communication experiment was the third opportunity for the country, after ATS-F and Symphony, laying a good foundation for satellite communication.

**Fig. 3.4** APPLE with Meteosat

# Chapter 4
# The Early 80s

## 4.1  INSAT-1

Indian National Satellite-1A (INSAT-1A) was built by Ford Aerospace Corporation of the US with Mr P P Kale as its Project Director. It was launched by American Delta Launch Vehicle on 10 April 1982 and prior to the launch, Master Control Facility (MCF) was established at Hassan for satellite operation and control. A resident team at Palo Alto, headed by Dr Vasantha, had coordinated the work. The team included Mr N K Malik (AOCS), Mr K Thyagarajan (System Engineering), S Y Ramakrishnan (power), Mr V K Kaila (Thermal), all from ISAC, Mr K L Valiappan (Propulsion) from LPSC, Mr K P M Bhat, Mr Damodaran from VSSC, Mrs Dipti Rastogi for communication payload, Mr Desai for VHRR payload, both from SAC. This was consistent with the ISRO policy of making satellite technology available to benefit society on an operational basis. INSAT-1 was a state-of-the-art multipurpose satellite with 12 C band transponders for communication, 2 S band transponders for relaying TV signals with India coverage and a Very High Resolution Radiometer (VHRR) giving a snapshot of the earth with 2.75 km resolution in visible and 11 km resolution in Infra-Red (IR), every half an hour. Also, there was a data relay transponder for receiving meteorological data in the 400 MHz band and disseminating in the S-band. It also had a Radio Networking and news gathering system. The mix of payload, particularly the VHRR, made this satellite a very complex one as this payload needed a radiation cooler to cool the IR detector to 105 K (minus 168 deg Celcius). This radiation cooler had to have a clear view towards the North, hence, no solar panel could be mounted on the north side. Thus, only one solar panel on the south side was put and a deployable solar sail was accommodated on the north panel to compensate for solar radiation pressure.

ISAC was not directly involved in the INSAT-1 project. However, a number of engineers from ISAC were on deputation and this provided close interaction. Professor Rao was chairman of the Project Management Board. INSAT-1A launched on 10 April 1982, failed on 4 September 1982, in its first eclipse after 4 months of successful operations and this was a setback to the communication satellite program

© Indian National Academy of Engineering 2022

P. S. Goel, *Making of a Satellite Centre*,

https://doi.org/10.1007/978-981-16-3480-2_4

of ISRO. This perhaps put some focus on APPLE, though not really as a substitute because APPLE had only two transponders in the C band. But as a capability demonstrator that we could successfully build and launch a geostationary communication satellite, its importance increased because of INSAT-1A failure. A Failure Analysis Committee was constituted with Prof Rao as chairman. I too was a member. The committee had limited access to design details as those could not be revealed to non-US citizens under the non-discloser agreement. Only an American consultant to the INSAT-1 project could see and discuss with design teams and later confirm to us that the design was good enough or they had verified from records that manufacturing was in order.

We created our own failure scenarios and forced designers to defend the design and come up with clarifications. This way, we could obtain sufficient insight into the failure mechanism and make sure that INSAT-1B under fabrication and testing would not have those design problems. Surprisingly, not much analysis could come out of our resident team because they were not allowed to have access to design details due to technology transfer clauses. It was also true that some members of the team learned a lot through this interaction that was useful in our INSAT-2 design and development. Mr Nageshwara Rao, deputed for subsequent satellite, got a very good understanding of solar sail and boom design, and Mr Malik got good exposure to the design of momentum management controller for V-mode and L-mode control law for high stability requirement for VHRR. Mr Thyagarajan got a good understanding of systems engineering. Mr Desai, Mr C M Nagarani and Mr Kaduskar, deputed for later satellites, developed a good understanding of the VHRR payload. All their experiences were gainfully utilised for INSAT-2.

INSAT-1B launched a year later on 30 August 1983 served very well for almost 10 years and revolutionised the TV broadcasting and communication scenario in India. By the end of the decade, almost 90% of the population and 80% of the Indian landmass of the country was covered by the Television through INSAT-1B. The methodology was to receive the weak INSAT-1B TV signals by Doordarshan in S band and retransmit them in the VHF band for local distribution. The use of VHRR for deriving the wind vector through cloud motion was operationalised. However, India Meteorology Department (IMD) was using all observations for the weather forecast in a synoptic manner, plotting this information and using human expertise to infer meteorology forecast for almost the next 25 years. Full exploitation of the VHRR data in the numerical model had to wait till 2008, when a new era of IMD modernisation began. INSAT-1C was launched in July 1988 on Ariane vehicle, but failed soon after. INSAT-1D was launched in June 1990 by Delta launch vehicle. These two satellites INSAT-1B and INSAT-1D served well through the late 80s and 90s.

## 4.2   SLV-3, ASLV and Space Science

While the satellite program was being driven by applications, the launch vehicle development with VSSC as the lead centre was moving at its own pace. ISAC was building satellite payloads for these flights. Mr Tarsem Singh was Project Director for the Rohini series of satellites. The first experimental flight of SLV-3E1 on 10 August 1979 was not successful, followed by the second experimental flight E2 on 18 July 1980 which was partially successful. Both had technology payloads. The first development flight of SLV-3, D-1, launched the RS-D1 satellite into designated orbit on 31 May 1981. The second development flight of SLV-3, D2, took place on 17 April 1983 and the RS-D2 had a smart camera developed by a team led by Dr Alex, and had a similar resolution as that of Bhaskara satellites (Fig. 4.1). As the application-driven program demanded heavier satellites and 40 kg was too small, it was decided to discontinue the SLV-3 and augment the launch vehicle by adding two first-stage motors as strap-ons and taking the capability to 150 kg. This new vehicle was called Augmented Satellite Launch Vehicle (ASLV) and the resulting satellite series was named Stretched Rohini Satellite Series (SROSS). Mr V A Thomas, Mr U N Das and Mr S M Bedekar were key members of the SROSS team.

The first launch of the ASLV D-1 in March 1987 was not successful as the first stage of the vehicle after strap-ons burn did not ignite. The second development flight of ASLVD-2 also resulted in failure in July 1988 due to heavy upper atmosphere winds and aero-elastic failure of the upper portion containing the equipment and the satellite. The third development flight, ASLV D-3 on 20 May 1992 resulted in partial success with lower orbit and the satellite SROSS-C re-entered the atmosphere after 2 months. Mr Das was appointed PD SROSS C-2 and Mr Bedekar as its APD. The ASLV-D4 successfully launched the SROSS-C2 satellite into correct orbit on 4 May 1994. The SROSS series of satellites were designed as spinning satellites carrying scientific payloads and having magnetic attitude control for spin rate and spin axis

Rohini Satellite (1982-83)          SROSS satellite - Integration (1986-87)

**Fig. 4.1**   RS D-2 and SROSS-C

orientation. It had six deployed solar panels on a hexagonal cylinder and six body-mounted panels. The SROSS-C2 had Gamma-Ray Burst Payload and a Retarding Potential Analyser payload. SROSS-C2 also had a monopropellant propulsion system (Fig. 4.1). By this time, PSLV had made good progress and ASLV was also too small for emerging polar missions, it was decided to discontinue ASLV and consequently the SROSS series of satellites. Though ASLV was not successful as a program, it provided a thorough understanding of launch vehicle technology and engineering through its aerodynamically unstable configuration. The great success of PSLV owes a lot to the lessons learned from the failures of ASLVs.

## 4.3   PTU to LPSC Bangalore

A collaboration between ISRO and a French company, SEP, working for Ariane, the European Launch Vehicle, resulted in the setting up of a unit in Bangalore by ISRO for manufacturing of Pressure Transducers under licence, named Pressure Transducers Unit (PTU), in the mid-seventies. The initiative was from Dr A E Muthunayagam and Mr V Srikantan was appointed as its head. PTU was a technical and commercial success and delivered a large number of pressure transducers for the Ariane program and also contributed to ISRO launch vehicles. Soon it became evident that liquid propulsion would play an important role in satellites; Prof Dhawan in consultation with Dr Muthunayagam decided to convert PTU into Auxiliary Propulsion System Unit (APSU), addressing the needs of spacecraft propulsion. Dr K Anantharam was appointed as its head. There was a strong suggestion from ISAC that APSU should become part of ISAC as it will primarily cater to Satellite Propulsion. Right or wrong, myself and Mr M G Chandrasekhar, with tacit support from Prof U R Rao, were arguing for APSU to be part of ISAC. Professor Rao could not openly argue as he wanted to be seen as more of an ISRO person. We were seen as ISAC loyalist young-sters. However, to consolidate the liquid propulsion activity, APSU was retained as a unit of the Liquid Propulsion Group that Dr Muthunayagam was creating at Valia-mala. APSU initiated the development of all elements of the Hydrazine-based mono propulsion system inclusive of a One-Newton thruster for forthcoming IRS satellites. In those fast-expanding days of activities, these were seen as opportunities to create separate functional units and to become independent directors. Acquisition of huge land at Valiamala hills, about 40 Kms from Trivandrum in a different direction (from VSSC), was partly to create a separate centre and partly a necessity to find enough space for setting up test laboratories/facilities for testing liquid propulsion systems under development. Dr Muthunayagam was spearheading the activity though still formally under VSSC.

## 4.4  PSLV and ISAC

The Polar Satellite Launch Vehicle (PSLV) project was formed in the early eighties with Dr S Srinivasan as Project Director to cater to the needs of launching Remote Sensing satellites into polar sun-synchronous orbit. I was a member of Project Management Board (PMB) to represent users from ISAC. It was a new opportunity for me to understand launch vehicle systems that I lost on joining SSD but with a different perspective. The PSLV project gave a new boost to liquid propulsion activity as its second stage L40 had a liquid engine. VIKAS, the liquid Engine, was an important and enabling component. Viking engine was being developed by Arianespace in collaboration with SEP of France. There appeared to be a shortage of skilled manpower in France for detailed designing of propulsion components. ISRO agreed to place its skilled manpower with CNES/SEP (French Space agencies) to augment its design efforts. It was very unique cooperation to provide engineering manpower support to a French company and in the process to acquire technology if we had the ability to understand and absorb. Mr Nambinarayanan, a senior engineer with Dr Muthunayagam, was given overall responsibility and he actually led the resident team of ISRO engineers for a few years at SEP France. On return of the team after the expiry of the contract, the team successfully developed the Vikas Engine and the L-40 s stage, which was a major step in developing indigenous capability for launch vehicles in general and for PSLV in particular. It was as important a milestone in Indian Launch Vehicle history as the development of solid rocket motors for SLV-3/ASLV under the guidance of Mr M R Kurup and Dr V R Gowarikar and that led to the development of the first stage of PSLV, the S-125 (now S-139) stage. We made sure that PSLV had a minimum launch weight capability of 800 kg in 800 km polar sun-synchronous orbit, a heat shield envelope of 3.5 m and sufficient mission flexibility to accommodate varying mission needs. Our contribution from ISAC was to give mission requirement definitions to PSLV. We introduced a salvage mission mode in case of a significant reduction in the performance during the launch. It will target another lower sun-synchronous orbit, based on the velocity to be gained algorithm. This capability was, however, never required to be used, though it exists even today.

PSLV has been the backbone of the remote sensing program of ISAC/ISRO. The first launch in 1993 with IRS-1E and Prof Rao as chairman was not successful and that was a big setback as Prof Rao had put his heart and soul into its development. The problem was quickly understood and subsequent record of successful launches with Dr Kasturirangan as chairman can be of envy of any space agency. The third and last development flight carried an IRS-P3 satellite having a unique combination of payloads and an opportunity for inertial referencing using gyros updated by star sensors. While IRS 1C was launched by a Russian launch vehicle, PSLV became operational for subsequent launches.

Dr Srinivasan, PD PSLV, wanted more space and moved the PSLV project to Valiamala, hence, the Valiamala campus was divided into two parts, Liquid Propulsion and PSLV, much to the discomfort of Dr A E Muthunayagam.

Due to persistent follow-up from Dr Muthunayagam, liquid propulsion activity was recognised as a separate centre and Liquid Propulsion System Centre (LPSC) was created, retaining APSU as a unit of LPSC, named LPSC Bangalore and Dr Anantharam as Group Director. Dr Muthunayagam later acquired land near Nagarcoil (his hometown) for testing big liquid rocket motors at Mahendragiri as another unit of LPSC. In 2011, this has been designated as a separate centre as ISRO Propulsion Complex (IPRC), and the contributions of this unit in preparing Liquid stages has been very professional and important. Though an independent centre, IPRC has worked very coherently with LPSC.

Understanding of this evolving system is an important part of the evolving culture of ISRO. Why is it that ISRO has succeeded more than other similar organisations in the country? Personal aspirations, loyalty to the Unit you are working for, are embedded in human genes and ISRO culture does allow to put forward individual views, but, once the ISRO leadership decides on a particular issue, all the differences are forgotten and everyone works for a larger goal. PSLV remained a project of VSSC and later, Dr Srinivasan became director of VSSC. LPSC Bangalore continues to be a unit of LPSC. The interaction between LPSC Bangalore and ISAC, functioning of matrix system with inter-centre coordinating mechanisms and overall supervision of Headquarters makes ISRO different. In ISRO culture, the best management decision is not that important as the culture makes it work, even otherwise.

# Chapter 5
# The Operational Space Services Era

## 5.1 The Remote Sensing Satellites

Bhaskara-1 and Bhaskara-2 provided invaluable experience in developing a methodology of satellite-based remote sensing, setting up ground systems for data reception, data archival, data pre-processing and data products to the user. Being experimental missions with a lower resolution of almost a kilometre, it was imperative that an operational mission with much better resolution had to be planned. A study team was constituted in the early eighties. Dr George Joseph was studying various aspects of payload and I was chairing the evolution of a new spacecraft bus for the operational series of Indian remote sensing satellite, IRS-1.

Professor Rao organised a familiarisation tour to the Soviet Union to understand the science and art of satellite remote sensing as the Soviet Union had a very comprehensive program, next only to LANDSAT of NASA. Dr George Joseph, Dr Ravi Dutt Sharma, then scientific secretary and I were deputed. We were taken to some laboratories working on processing satellite data but were not allowed to see any payload or satellite hardware.

In the year 1981, the project IRS-1 was created and Dr Kasturirangan was designated as Project Director. Our baseline for the payload was LANDSAT-2 and 3 as we in India had developed sufficient expertise in interpreting and using the LANDSAT data. We were also keeping an eye on the development of French satellite SPOT in our planning as it offered better resolution. A four-band Payload, three in visible and one in near IR, was suggested. So far, all remote sensing satellites in orbit or in planning were using scanning mode of detection using a single detector for each band. The area coverage is by the relative satellite motion of the ground trace by about 7.8 km/s. The team at SAC came up with a new detector under development, Charge Coupled Device (CCD) by Fairchild, a US company and this was a landmark decision as this not only simplified the development but also, later, put ISRO ahead of advanced countries in the arena of satellite remote sensing. Fairchild was happy

© Indian National Academy of Engineering 2022
P. S. Goel, *Making of a Satellite Centre*,
https://doi.org/10.1007/978-981-16-3480-2_5

to give these detectors so as to trigger the entry of their technology into the remote sensing satellite market. The optical telescope had to be imported, tailored to our design.

The new spacecraft bus had to be 3-axis stabilised with earth pointing payload in a sun-synchronous polar orbit. We had the experience of APPLE with momentum-biased 3-axis stabilisation. However, this will pose a limitation of roll biasing and is associated with nutation, which might pose a problem to the high stability requirement of the payload. The short-term stability requirement was $3 \times 10^{-3}$ degree/sec so that the images do not suffer in quality due to smear. Hence, a zero momentum-biased, 4 reaction wheel-based attitude control system was selected. Three reaction wheels, with one each mounted along each axis, were the baseline and the fourth was skewed along the resultant axis. It was providing redundancy to take care of any one reaction wheel failure. We did not have long-life gyros of our own, but Dynamically Tuned Gyro(DTG) was under development at ISRO Inertial Systems Unit (IISU), Thiruvananthapuram, we decided to control pitch and roll from scanning Earth Sensor and yaw from redundant DTG. We imported a few DTGs from SAGEM France as backup but used our own, developed by IISU in IRS-1A. The performance of our DTG was very good and the whole credit goes to very systematic development by Dr Krishnan at IISU.

The story goes back to 1982 when a committee was appointed to review the development of Gyros in ISRO/IISU. A floated Gyro developed for SLV-3 vis-à-vis the DTG was under development for satellites and Launch Vehicle missions. Dr S C Gupta chaired the committee and I was representing ISAC and it was clear to us that Floated Gyro would not meet the long life and stability requirement of satellites, though VSSC opted to continue with floated Gyros for the launch vehicles. Hence, both developments were pursued, but our thrust was on DTG. IRS-1A in a way changed this and DTG became a workhorse for ISRO, both for satellites and launch vehicles for subsequent three decades. Dr Krishnan continued to work on improving DTG performance and reducing its weight throughout his long career in IISU. ISRO/ISAC owes a lot to his dedicated effort and perseverance.

The story of IRS is one of leapfrogging in technology in many areas: Zero Momentum 3-axis Attitude Control system, DTG, Scanning Horizon Earth Sensors, Solar Array Drive, Reaction wheels, Hydrazine-based monopropellant system including 1 N Hydrazine thrusters, bladder tanks, flow control valves, latch valves with related testing and qualification, TTC shifted to S band, 20 MBPS data transmission in X band using QPSK modulation, data handling, a backup option of 5 MBPS in S band, interfacing a 12-channel analog tape recorder imported from the US, a new spacecraft bus that is serving even today even after three decades, and of course a CCD-based imaging payload that provided a worldwide lead to ISRO in contemporary remote sensing scenario. Each of these developments has a hero or a group behind it.

With the launch of IRS-1A on 17 March 1988, ISRO became a serious player and India became a leading country in remote sensing in the world for decades. IRS-1A development was spread over 6 years as everything was a new development. The control logics were complex, designed using CMOS SSI/MSI technology except for a few functions like a 13-order Kalman Filter, modelling orbit frequency (and harmonics) attitude biases and reaction wheel auto reconfiguration against wheel degradation, which were implemented using an RCA 1802 microprocessor. Dr Murugesan was leading control electronics development from Bhaskara onwards and was joined by Dr VK Agrawal who played a key role in later years. Dr Murugesan left ISAC in the mid-80s for a professorship in Australia and Dr Agrawal carried AOCE development forward.

IRS-1A was a major step and hence was another opportunity for ISAC to grow. We added about 200 engineers/scientists, a large number of facilities like upgradation of computing facility, addition of test facilities, new cleanroom, 3-axis servo table and expansion of all laboratories. During this period, we moved to a new building adjacent to the NAL campus, Kodihalli, on HAL airport road in October 1984. A large cleanroom was set up along with mechanical integration facilities adjacent to Ground Checkout facilities. Each subsystem was allotted a separate lab in the new building. It was a big change for electronics designers, with each subsystem getting a separate lab as against just two work tables in a common lab in Peenya. An indigenous modular anechoic chamber was built for comprehensive Electromagnetic Interference and Electromagnetic Compatibility (EMI/EMC) testing. A big facility came for wiring and assembly of electronics packages, the Central Electronics Fabrication Facility (CEFF). Mr Gopalakrishna, a senior tradesman, organised the CEFF so professionally that it became a showpiece of ISAC for decades to come.

A state-of-the-art bonded store for space-qualified components was created. QA created a Test and Evaluation (T&E) laboratory near the cleanroom so that a package could directly be delivered to the cleanroom after the T&E. QA also set up a modern facility for component-level failure analysis. The Environmental test facilities were augmented with a 4 m thermo-vacuum chamber, multiple numbers of 1 m chambers and dozens of hot and cold chambers. The 2 m thermo-vacuum chamber set up under the APPLE project was relocated to the new campus. An eight-ton vibration shaker was added to test heavier IRS satellites. An anechoic chamber was built using indigenous materials by Mr V R Katti and Dr S V K Sastry for EMC testing of electronic systems and a separate EMC test facility was set up.

IRS-1A was the result of a unique combination of management and technology evolution under the Project Directorship of Dr K Kasturirangan and the overall guidance of Prof U R Rao. The core team under Deputy Project Director (DPD) Mr S Kalyanaraman was meticulously perusing, following up each and every action item. Dr M G Chandrasekhar, as DPD mission, evolved complete mission management structure from spacecraft operations to data processing, data dissemination and utilisation, well supported by Mr S K Shiv Kumar. Dr Surendra Pal and his team designed the S band TTC and X band RF systems. He came up with an interesting concept of compensating higher range with higher gain and reducing antenna gain for near nadir position of the ground station. Mr R K Rajangam TT&C designed the digital TM/TC system. A separate activity of payload data handling was created with Mr E Vasantha and Ms Nancy Nelson. Mr S G Basu designed a ranging system. Dr M S S Prabhu was responsible for spacecraft structure and Mr H Narayan Murthy and Mr D R Bhandari developed a thermal control system. Mr Y K Jain, Mr Kanaka Raju

under the guidance of Dr T K Alex developed the Scanning Earth Sensor, Analog
Yaw Sun Sensor and other attitude sensors.

The payload was a teamwork with Dr George Joseph as DPD payload, well
supported by Dr Ram Ratan (optics), Mr Kiran Kumar (detectors and overall system
engineering) and Mr Nagachinchaiah (electronics). This trio under the overall leadership of Dr George Joseph developed the best opto-electronics payloads for ISRO
satellites in the world in the subsequent years and established ISRO in the place
we find it today in remote sensing technology. Mr N Vedhachalam was DPD inertial
systems in addition to being director of IISU and was responsible for the development
of DTG, Reaction Wheels and Solar Array Drive. Dr Ananth Ram was responsible
for the complete mono propellant system in addition to being head of the Bangalore
unit of LPSC. Mr A D Dharma with Mr K S V Seshadri and Mr O P Sapra carried out
spacecraft assembly, integration and testing activities with uniquely designed spacecraft and payload checkout systems. The methodology to evaluate complex optical
payload with high data rate was established.

My role as Chairman Spacecraft Systems Advisory Board (SSAB), in addition to
being Head, Control System Division, was to tie up all the new developments and
assiduously follow up. This was in addition to design and development of AOCS. It
required a total system engineering approach and understanding of the criticality of
individual subsystem development simultaneously.

IRS-1A (Fig. 5.1) was the first operational class of satellite that we were building
and also the first where all basic building blocks of the satellite were being developed.
Inertial elements like Gyros, Reaction Wheels and Solar Array Drive, Monopropellant System with components like 1 N thrusters, flow control and many other valves,
Scanning Earth Sensors and high bit rate data handling system were all new. In addition, we were also building test facilities, operational ground systems and creating
a robust space culture with space quality systems, testing methodologies and standards. IRS-1A data was received at Shadnagar (Hyderabad) ground station of NRSA
which was till then used for LANDSAT and Bhaskara's data reception. The ground
station was upgraded by Mr D V Raju who had moved from ISAC to NRSA earlier.

**Fig. 5.1** Schematic of IRS-1A in orbit

Professor Deekshitulu was Director of NRSA. This mission was to put ISRO into the operational era of remote sensing. Dr Kamat at SAC had evolved the remote sensing data utilisation framework for Bhaskara-1 and 2 which came very handy now to put an operational system in place.

## 5.2 A Permanent Home at NAL Campus

Amongst all the excitement of these developments, ISAC was yet to get a permanent building as its home. Early in 1980, Prof Dhawan convinced National Aerospace Laboratory (NAL) to give part of its space consisting of a Guest House and vacant land to ISRO for creating a building for ISAC and NAL initially parted with about 13 acres of its area. Professor Rao was very keen to start the building and a private contractor was engaged as per norms. The contractor after completion of the foundation stopped the work demanding more money, perhaps a normal practice in state government. This was not accepted and a GoI Company, viz., National Building Construction Corporation (NBCC) was entrusted to take up the construction of the building.

It is around this time that an old request of Prof Rao to the government of Karnataka fructified and 400-acre land was allotted by the state government at Yelahanka. It was a golden opportunity to create ISAC as an integrated complex including housing. Some of us pleaded to accept and change plans, but perhaps it was too late as the commitment to build at NAL campus was already made or perhaps it would look like going back on personal pursuance of Prof Dhawan with NAL. Yelahanka was considered too far at that time. We will never know the exact reason, but a good opportunity was not exercised and lost.

We moved to the new building during August–September, 1984, still retaining some of the sheds in Peenya, which were later used for setting up Laboratory for Electro Optical Systems (LEOS). Professor Rao's stay in this new building was for only a few days (which he built with enormous passion) as he was appointed as Chairman ISRO from 1 October 1984 and moved to Anthariksh Bhawan, at New BEL Road, which had just been constructed and also ISRO HQ had just been shifted from Cauvery Bhawan only a few months back.

## 5.3 IRS-1A Mission

IRS-1A was perfectly launched on 17 March 1988 from Baikanur by a Soviet-built Intercosmos launch vehicle. ISTRAC had shifted to Peenya, Bangalore, meanwhile and Mr K V Venkatachary, Director ISTRAC, had created a new ground station and control room. We acquired the first signal from Lucknow station and verified that the spacecraft was in good shape and solar panels had automatically deployed after separation from the launch vehicle. Many satellites elsewhere had solar panel deployment problems in the orbit in other space missions, the primary cause being thermal gradients across the panel and supporting hinges. Normal practice is to heat the panel by orienting it towards the sun and minimise the gradients. This is what we also do for Geostationary Transfer Orbit (GTO) launches as the panel deployment can only take place after the Liquid Apogee Motor (LAM) firing. However, we decided not to let thermal gradients develop and deploy the panels soon after the launch. The strategy worked well and we had a proper deployment of both the solar panels soon after launch. This strategy is being followed as a standard practice by ISRO till today for all Low Earth Orbit (LEO) missions.

Immediate operation after the solar panel deployment in orbit was to have earth acquisition in pitch and roll axes through Scanning Earth Sensor and capturing yaw axis through Analog Yaw Sun Sensor. In polar sun-synchronous orbit, it is noted that this sensor gives correct yaw angle only at the poles and yaw control had to be transferred to the gyro-based reference after resetting the gyro output at poles. It had to be done automatically at the South Pole as we had no commanding capability near poles. This auto action was to be verified only in the next visible orbit. We had perfect yaw acquisition and proper update at the North Pole and the next pass activity was to transfer controls to reaction wheels for fuel saving as the spacecraft was getting into non-visibility for the next 3–4 orbits. Subsequent major activity was to stabilise thermal control system by bringing in appropriate heaters and do minor orbit corrections to put the satellite into the proper polar sun-synchronous orbit and prepare the satellite for payload operations after one week of outgassing of X band TWTs although initial data was received through a solid-state (5 MBPS) amplifier. The first payload operation was commanded from Lucknow station and the first pictures of landmass were received at NRSA, Hyderabad, from Mansarovar to Rameshwaram. We all became ecstatic looking at the images on quick look display. India had entered the operational Remote Sensing era!

IRS-1A has not only been a major milestone in the journey of ISAC but has many stories associated with it that are part of ISAC's evolution. A study team was constituted under the chairmanship of Dr Kasturirangan to evolve IRS spacecraft. I was asked to chair the spacecraft bus configuration to accommodate the remote sensing payload, being evolved under the guidance of Dr George Joseph. Since we did not have baseline data on any of the packages, viz., power, battery, TM/TC, AOCS, sensors and actuators, we had to start with the best guess for every package. We decided to choose S band for TTC and X band for data transfer to meet the operational needs of high remote sensing data rate. We decided to develop Hydrazine-based monopropellant RCS to be developed at LPSC Bangalore. The 10 NMS reaction wheels, Dynamically Tuned Gyros (DTG) and Solar Array Drive (SADA) had to be developed at the newly created Inertial Systems Unit (IISU) at Trivandrum. Scanning Earth Sensors and many sun sensors like Analog Yaw Sun Sensor had to be developed at Sensors division headed by Dr Alex.

The zero momentum 3-axis control system was to be developed by us in the control systems division at ISAC. Each of these activities has its own story of setbacks and successes during the subsequent 4–5 years. Myself as Chairman Spacecraft Systems Advisory Board (SSAB) and Mr Kalyanaraman as Deputy Project Director were following the related developments each day and resolving technical and interface issues. The RCS thruster degrading or a latch valve leaking under test, DTG flexure breaking or the SADA getting stuck and many such problems were addressed meticulously and seriously. This exercise set a mechanism that we now call "team ISRO". IRS-1A was the first such exercise in which we learnt to work as an interdisciplinary and inter-centre team ISRO.

In 1986, we had a big setback. Dr Kasturirangan was taking payload review as Project Director at SAC. The meeting started at 9am in the morning and went up to 7 pm, as usual. I was sitting by the side of Dr Kasturirangan and Dr George Joseph and Mr Kalyanaraman were sitting on the other side. At the end of the meeting, I noticed that Dr Kasturirangan was not able to pick up his pen. As I alerted Dr George Joseph, suddenly, Dr Kasturirangan collapsed in the chair. It was a massive paralytic stroke. SAC immediately organised to shift him to a hospital and was well attended. He returned to Bangalore after about 2 weeks of treatment and some rest. What is remarkable is that he very quickly took control of himself in terms of medication, exercise and IRS-1A project in such a way that the serious ailment just became a small event. What determination and will power? It never reflected again in his long professional career till date, even after three decades and pressures and pulls of ISRO chairmanship, Membership of Parliament, membership of planning commission and many other responsibilities.

The successful launch and initial operations of IRS-1A was just an enabler and the real objective was to develop applications and make them operational. SAC and NRSA did such an excellent work involving Regional Remote Sensing Centres, government departments like agriculture, water and mining, that IRS became a symbol of national development and growth. A large number of private companies sprung in and around Hyderabad to assist in developing applications.

Indian Space Program became synonymous with space for national development, a unique identity that we are all proud of and what the founder (Prof Sarabhai) had said in UN some 20 years earlier (20 years with reference to IRS-1A launch in 1988). Professor Dhawan had created a unique mechanism of utilising the remote sensing data across various ministries, named as National Natural Resource Management System (NNRMS). It consisted of 13 committees, each chaired by the respective Secretary of the department and served by a secretariat at ISRO HQ. Today, we are a leading nation in the world as far as remote sensing applications of space are concerned and IRS-1A is the landmark mission in this context (Fig. 5.2). With it, the Indian Remote Sensing era had begun. The list of ISRO remote sensing satellites is very long and we may subsequently touch upon some interesting developments in this domain.

## 5.4  The Satellite Communication

We had APPLE in June 1981 for the effective demonstration of the capability to build a communication satellite and prove related technologies and ground systems.

**Fig. 5.2**   IRS Remote Sensing Applications

APPLE, as an experimental forerunner to INSAT, provided an opportunity to experiment with various applications like VSAT, TV service, etc. INSAT-1A had initiated the development of ground infrastructure and that started growing at an unprecedented rate after the launch of INSAT-1B. Practically, one or the other Doordarshan receive/transmit system was being inaugurated every day and the communication services were fast expanding. VHRR pictures were shown along with news bulletins on the weather. While SITE was an experiment for the unprivileged in certain select regions, INSAT-1B was bringing satellite communications to the masses. Professor Dhawan created another body INSAT Coordination Committee (ICC), involving user ministries like the department of Telecommunication, Wireless Advisor, Doordarshan, etc., to provide coordination between satellite provider ISRO, service providers like DoT and Doordarshan and the regulatory mechanism like International Telecommunication Union (ITU). This mechanism has well served the country for over three decades and more. Important services like e-banking, National Stock Exchange and e-governance started to utilise satellite-based communication effectively.

INSAT-1 brought new inspiration to society as a whole. While Aryabhata was an inspiration in terms of technological pursuit in India, INSAT-1 brought about a change in the mindset towards digital connectivity. Figure 5.3 shows the INSAT-1 communication network.

APPLE experience and INSAT-1B impact were the driving force to undertake our own INSAT-2 program in the mid-80s. Mr P Ramachandran was appointed to

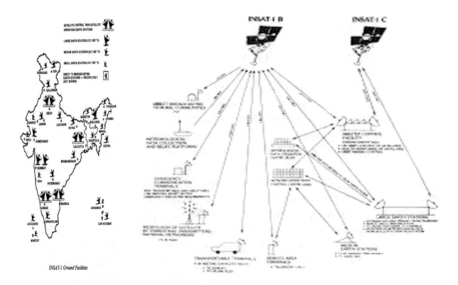

**Fig. 5.3**  INSAT-1 communication network

head the study team for the development of indigenous communication/multipurpose satellite systems. He came up with a payload configuration of INSAT-2A and 2B with 12 C band transponders, 6 Extended C band transponders for communication, 2 S band transponders for TV broadcast and an improved Very High Resolution Radiometer (VHRR) for meteorology in consultation with SAC and users. Interestingly, we consulted the communication user segment, DoT in particular, to understand the projections for the late 90s so as to plan for the decade and were told that the capacity of INSAT-2A and 2B would more than suffice for the decade of 1990s. How wrong these assessments were! The satellites' capacity was fully booked as soon as these satellites were launched. The utilisation growth of these satellites was much beyond projections and this is the primary cause of perennial satellite transponder shortages.

I had chaired the spacecraft configuration committee for the new spacecraft bus for INSAT-2. It was once again a central cylinder-based configuration, cylinder supporting oxidiser tank inside and two fuel tanks and other propulsion components outside the cylinder. We decided to have an indigenous bipropellant propellant system, indigenous momentum wheels, SADA and Earth Sensors. Complete payload including complex VHRR was to be developed at SAC. However, critical passive radiation cooler and scan mechanisms for VHRR payload were to be developed at ISAC and IISU, respectively, making it an inter-centre developmental effort. While configuring INSAT-2, we had the constraints of VHRR passive cooler to look into space, so as to cool the detector in the range of 105 deg K. As the earth is nearer to Sun in the winter solstice, about 3% higher solar radiation would be there on the south face in winter compared to the north face in the summer, the cooler had to be North facing. Consequently, a single solar panel had to be on the south face, making the satellite asymmetric. INSAT-1A and 1B had similar configuration, hence, we were familiar with the issue of asymmetry and its impact on slow attitude drift. The V-configuration corrects for role error while yaw error gets converted into roll error in quarter orbit.

INSAT-2 was a major program following IRS-1A, a big project in terms of infrastructure and manpower augmentation. Facilities like Large (9 m) Space Simulation Chamber (thermo-vacuum chamber with 4 m diametre sun simulator), LSSC, 16-ton vibration table, large cleanroom, 3-axis servo table, a large computer facility and a new INSAT building were added. About one-third of the current manpower of ISAC was added under INSAT-2 and there has been practically no addition of manpower since then. In a way, this was the completion of the Satellite Centre in the making since the Aryabhata project, a journey of a decade and a half.

Professor U R Rao was overviewing the INSAT-2 project study, planning for facilities and augmentation of the manpower as director ISAC in the year 1984 and had just entered the permanent building at NAL campus (present location at HAL airport road) during August–September. He then shifted to ISRO HQ on October 1 as Chairman ISRO on the superannuation of Prof Dhawan. His dream, perhaps the most sought-after goal or purpose of his life, was achieved with ISRO Satellite Centre getting a permanent building. It was time to extend his own personal aspirations beyond ISAC. But, whether as chairman ISRO for the next 10 years or a mentor at ISRO HQ later, playing many key roles at the international forum as vice president of International Astronautical Federation (IAF), guiding space science as ADCOS chairman, chairman of Prasara Bharati or member of Reserve Bank of India Board, his heart was always at ISAC. ISAC was his creation, his child and his spiritual home.

Colonel Neelambar Pant (Pant Saab as we called him) was shifted as Director ISAC from Shrihari Kota, SHAR. His shifting from SHAR was marked by nature as a very unusual, unprecedented event occurring in the form of a super cyclone. In October 1984, this super cyclone not just struck at SHAR, but remained static over SHAR for 3 days, playing havoc, cutting off SHAR from the mainland. There were many stories of suffering and bravery by stranded ISRO staff locked up, totally isolated with all communication systems disrupted. But one story is worth recollecting. Head EMD SHAR had come to Chennai for some work and was locked out. Since head EMD was in charge of all facilities, civic amenities, he felt he must be there at SHAR to help the director (Col. Pant) and the people. Mr S M Loganathan swam almost a kilometre to reach SHAR and took charge. This is the ISRO culture. Colonel Pant joined ISAC only after he settled SHAR to normalcy, restoring all services, after the super cyclone died down.

The INSAT-2 project report was sent to ISRO HQ for two satellites for getting the approval of the Government, but came back from Satellite Communication Program office with a comment that it could not be called INSA-2 as INSAT-2 had to be an operational satellite. Since we were making such a satellite for the first time, it could only be called as a "Test Satellite". The name was modified as INSAT-2 TS. It was a bit painful but left with no choice after a mouthful of arguments. The project approval came and the project team was constituted. Mr P Ramachandran was appointed as

Project Director, and three Associate Project Directors, me as APD for AOCS, electronics and propulsion, Mr M N Sathyanarayana for mechanical systems and Dr K N Shankara at SAC for communication payload. We had a Deputy Project Director for each major subsystem. Mr N Vedhachalam, the then Director ISRO Inertial Systems Unit (IISU), Trivandrum, was DPD for inertial systems including momentum wheels, DTG-based Inertial Reference Unit (IRU) and Solar Array Drive; Dr Anantharam was DPD propulsion for the development of bipropellant propulsion system and all propulsion elements have to be developed anew. Mr Malik was DPD Attitude and Orbit Control System (AOCS) to develop the V-configuration momentum biased system with very low drift of the spacecraft to take care of the stability requirement of the VHRR payload. Mr V R Katti was DPD Integration, Mr Y N Bhushan was DPD mission, Dr P S Nair DPD structures, Mr V K Kaila DPD thermal, M Nageshwararao DPD mechanisms and Mr R K Rajangam DPD Digital Systems, TTC, etc. Mr Bhaskaranarayana was DPD TTC (RF). Mr S T Venkataramanan was DPD for Power Electronics and Battery and Mr N Srinivasamurthy for solar array, Dr V S Iyengar was Deputy Project Director for VHRR at SAC. INSAT-2 TS project had a very strong core team with Mr D Rajagopalan as Project Manager, Mr K L Valliappan as Manager for Propulsion, Mr C R Srinivasan for Manager Communication payload and a few others.

Though started a few years later than IRS-1A, the two projects were under development concurrently till the launch of IRS-1A in March 1988. The systems that had to be evolved anew in the spacecraft bus for the GEO mission primarily were Thermal Control System, Attitude Control System, C band TTC, antenna deployment mechanism, solar array deployment mechanism, sail and boom deployment mechanism and Power System. While there is an advantage of continuous visibility from GSO, longer exposure of the same area under the Sun poses temperature control issues. The satellite integration of this class with a complex payload mix was a big challenge and communication payload testing needed a dedicated checkout system to be developed. The whole checkout system was in auto mode with the flexibility to test or skip any module.

Colonel Pant had wide experience in communication systems, having designed the country's first communication Earth Station at ARVI (Maharashtra), had good administrative experience at SHAR and was devoted to the development of space systems with great zeal and commitment.

Meanwhile, Dr S P Kosta left ISAC to become director of SAMEER, an autonomous laboratory under Department of Electronics. His contributions in the making of ISAC had been twofold. First, he mentored developing antenna design capability, in particular Omni antenna for TTC and mentored Dr S Pal and his team. Second, and much more in terms of making the satellite centre, he was number two to Prof U R Rao acting as a cushion between him and the rest of the staff in technical and administrative domains. He was very easy and cordial with people at all levels and anyone could go to him and express his/her feelings. Dr Kosta was very tactfully implementing decisions of Prof Rao palatable or unpalatable without causing any unpleasantness. This was a very important role in the formative years of the seventies and eighties.

Mr Vasagam went back to VSSC as Deputy Director Avionics and later left ISRO to take over as vice chancellor of Anna University, Chennai. Mr Vasagam's unique contribution was APPLE, from concept to completion. He was one of the few, mostly found in the library, and had contemporary information on satellite programs in the world. Many persons called him a moving encyclopaedia. He was also Group Director, Control Systems while PD APPLE.

This period of eighties in which we had overlapping IRS-1A and INSAT-2 TS projects was the real period of making the Satellite Centre a happening centre. We matured as a Matrix Organisation Structure and most technology developments taking place in technology divisions and projects had to be coordinated through project core teams. A very formal and formidable Reliability and Quality Assurance team was organised by Mr Ramgopal. Mr Siharan De emerged as component expert and he would find a way to get any desired component of the requisite quality. Mr De was so popular with Post and Telegraph department that a post from the US correctly landed on his desk with just addressed as "Mr Siharan De, India".

Dr Gambhir, an alumnus of IISc built the state-of-the-art environmental facilities while nurturing Mr B N Baliga as a worthy successor. Thermal Group was headed by Mr H Narayanamurthy. He organised the three segments, (i) Thermal Analysis and Design, (ii) Testing and Evaluation and (iii) Thermal Control Implementation. Mr D R Bhandari, Mr P P Gupta and Mr Bhojraj emerged as key persons in the activity for the development and actual work on the satellites. Dr K Kasturirangan was heading the space physics program as head Technical Physics. He then took over as PD IRS-1 and later took over as Associate Director. A S Prakasa Rao was heading the Mission team.

Mr Tarsem Singh was looking after Rohini series of satellites as Program Director with Mr V A Thomas as Associate Program Director as well as Project Director and also Mr U N Das as Project Director for Rohini series of satellites. Mr A D Dharma was heading the Ground Checkout group with Mr O P Sapra in charge of IRS series and Mr V R Pratap for the INSAT series. Mr V R Katti was heading Spacecraft Integration and also was responsible for the INSAT series with Mr K S V Seshadri looking after IRS satellites. Mr R Ashiya was deputed to Paloalto as resident head for INSAT-1C and 1D as Dr Vasantha from VSSC took over as Project Director. Mr P P Kale went back to Ahmedabad as Director SAC. A large number of Technology Development Programmes (TDPs) were initiated to bridge the technology gaps. We were approaching towards making ISAC a "technology complete" satellite centre.

Dr Kasturirangan took over as director ISAC on 1 May 1990 as Pant Saab moved to ISRO HQ as Deputy Chairman ISRO.

Coming back to INSAT-2 TS, we had many technical and managerial challenges. Amongst the technology challenges, we had to start with the assumption that the spacecraft should be compatible with commercial launch vehicles from Europe, the USA and Russia. Since we had APPLE launched on Ariane and by this time it had evolved as a reliable launch vehicle at a competitive cost, we were familiar with all interface issues and launch pad operations and constraints; we optimised INSAT-2 interface to Ariane but the option to be compatible to other launch vehicles as well was retained.

We decided to leapfrog to a unified bipropellant system, a central cylinder-based load-bearing structure. We had to take up the development of 50 kg Apogee Boost Motor (ABM), 22 N thruster for attitude control and all the other elements of the propulsion system had to be designed and developed afresh. This was the first time that we decided to develop a redundant Inertial Reference System using three Dynamically Tuned Gyros developed at IISU, Thiruvananthapuram. IISU was also developing Momentum Wheels and a Reaction Wheel for V-mode of momentum biased attitude control system, in addition to all new and bigger Solar Array Drive.

As the telemetry data rate had to be increased to 1 KBPS, we decided to use payload band (C band) for TTC as well, all new development again. One of the key developments was Boom and Sail assembly to compensate for solar radiation pressure from a one-sided solar array and this remarkable work was done by Mr Nageshwara Rao, under the overall guidance of Mr M N Sathyanarayana. Dr Alex and his team, Mr Y K Jain, Mr Kamalakar, and Mr Koteshwara Rao developed Scanning Earth Sensor and other sun sensors. Mr V K Kaila and Dr R A Katti developed the thermal mathematical model of the satellite and designed the thermal control system.

INSAT-2 gave a beginning to designing, manufacturing and testing of Carbon Fibre parabolic antenna at the Composite facility of VSSC. Yet another important development was a passive cooler to cool the VHRR IR detector to 105 K by Mr P P Gupta and Mr S C Rustogi of the thermal group. The complex reflective coating was done at National Aerospace Laboratory (NAL). The polishing of these reflective panels was done at the sensor's lab at ISAC. IISU developed the scan mechanism for the VHRR to scan the earth for imaging. Development of C and S band communication payload and the complex VHRR payload was done at SAC under the overall guidance of Dr Iyengar and the main contributions came from Mr Nagrani, Mr Kaduskar, Mr Kiran Kumar and Mr Dave. Thus, many new technology developments were part of the INSAT-2 project at ISAC, SAC, IISU, LPSC, Composite group of VSSC and NAL, but all under the overall coordination of ISAC/INSAT-2 Project. The launch preparations at Kourou, French Guiana, were long, almost 2 months and the much awaited event happened on 10 July 1992.

The launch campaign at Kourou was long but eventless except that the momentum wheel designer M/s Teldix noted in-orbit anomalies in some of their momentum wheels in other satellites and decided to change lubrication methodology, hence proposed to change the wheels with a new qualified process. Rather than bringing the satellite back to Bangalore, it was decided to change the wheels at the launch complex with limited facilities and without going through validation tests like vibration or thermo-vacuum tests at the satellite level. It needed very careful handling of operations by the integration team and was meticulously executed.

The Ariane-4 launch vehicle lifted off majestically and injected the satellite INSAT-2A in a very precise orbit. Having overseen the pre-launch preparations, I had come to MCF Hassan to supervise the post-launch operations. One of the early activities after confirming the health of the satellite was to calibrate the Gyros to compensate for on-orbit drifts due to zero "g" and temperature variations. It was a complex operation as the yaw axis was locked to the Sun and the geometry was changing continuously. However, it went on well and provided a lot of confidence to the operations team. The first Liquid Apogee Motor (LAM) firing was planned for the third orbit, which was visible from Hassan. While preparing for the LAM firing, pressurised Helium gas was allowed to enter the fuel tank by firing a pyro valve and the gas entry was to be regulated by a pressure regulator at 17.5 bar. However, the tank pressure started going up as the pressure regulator was leaking. It was clear that if immediate action was not initiated, the tank with a burst pressure of 24 bar may burst, resulting in loss of the satellite. I initiated the LAM firing immediately. Luckily, the pressure regulator leak was within fuel flow and the pressure inside the tanks was controlled to within 22 bar. Now, we had to quickly decide the duration of the burn, much beyond the normal plan of about 60% as the latch valve had to be closed at the end of the burn. The pressurant in the tank should be enough to sustain subsequent burns and for lifelong operations. We decided to go to about 80% of the burn based on the back of envelope calculations. The strategy worked well and INSAT-2A (Fig. 5.4) was in orbit working satisfactorily.

INSAT-2B was launched a year later to augment the capacity and also as a backup. The tag Test Satellites (TS) was never used after the launch as these satellites provided very useful services to the nation as operational satellites. Post-launch observations, in particular two anomalies, are worth discussing. VHRR cooler had a temperature monitoring sensor which was used to control the detector patch temperature and to set temperature. Both the main and redundant sensors got detached after about 1 year. Their ground testing was done by dipping the mounted detector in liquid nitrogen, considered a much severe condition. However, months-long simulations showed that the sensor does get detached even at lower temperature excursions at a large number of cycles, not thought of before launch. The process was modified subsequently. Secondly, C band Solid State Power Amplifiers (SSPAs) started failing at the rate of one in a few months. This was never expected as solid-state devices are expected to be much more robust than Travelling Wave Tubes (TWTAs). Again a year-long study at ISAC, SAC and vendor's place indicated metal migration shorting the gate resulting in the failure. But in spite of these problems, INSAT-2A and INSAT -2B served their intended life and the satellite communication in India entered into a new phase of exponential expansion.

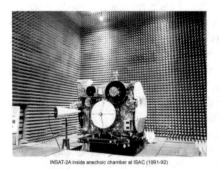
INSAT-2A inside anechoic chamber at ISAC (1991-92)

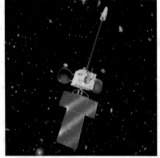

**Fig. 5.4**  INSAT-2A

# Chapter 6
# The Foundation TEAM, ISAC

As we discussed, Prof U R Rao was the chief architect of ISAC, conceiving it even before Aryabhata was initiated, and he used Aryabhata as an opportunity to move satellite activity to Bangalore, an important milestone towards setting up ISAC. However, there were many others who contributed to the cause in their own way largely by creating competence in their own area of expertise. Dr S P Kosta served as a sincere follower and faithful supporter, taking any odd jobs assigned to him by Prof U R Rao. He served as an interface between staff and director on all sensitive issues. He provided strength to Prof Rao in multiple ways. Dr Kasturirangan accompanied Prof Rao from PRL and provided space science support, member secretary to Aryabhata management board, heading technical physics division and interfacing with scientific community on behalf of ISAC/ISRO. Later as Project Director for Bhaskara-1, Bhaskara-2 and IRS-1A, he laid a strong foundation to remote sensing program. He was the only one to advise Prof Rao on all matters of strategy, technical or managerial.

By the time of launch of IRS-1A, Prof Rao had perhaps made up his mind that Dr Kasturirangan should not only succeed him as director (after Col. Pant) of ISAC but also as Chairman ISRO. He carefully crafted the succession plan through fast promotions to Dr Rangan. Dr Rangan succeeded Col. Pant as Director ISAC in the year 1990 after Pant Saab was appointed as Deputy Chairman, ISRO, and he focused on consolidating the centre's activities, ensuring success through review system and filling technology gaps. Colonel Pant gave a new direction to ISAC as director, of highest personal discipline, sincerity and integrity. While his discipline came from his background from army, his personal qualities were very different from that of a service man. He gave ISAC a fresh dimension of finest human touch and loyalty to the organisation. His own expertise on satellite communication played a role in connecting ISAC to application-driven thinking.

In terms of building ISAC as a strong satellite technology centre, there are hundreds of mid-level managers/engineers (at that time of 80s and 90s) who made ISAC what URSC (Fig. 6.1 shows some of them who participated in Bhaskara-1 launch from USSR cosmodrom. Professor Dhawan and Prof Rao are in the centre)

© Indian National Academy of Engineering 2022
P. S. Goel, *Making of a Satellite Centre*,
https://doi.org/10.1007/978-981-16-3480-2_6

" BHASKARA " team in U.S.S.R.

3

**Fig. 6.1** Prof Dhawan, Prof U R Rao, Dr Kasturirangan and others at the USSR for launch of Bhaskara-1

is today and some important contributions are described in the following sections, realising the fact that this is a difficult exercise and bound to have many omissions due to the inability to retain the reader's interest with technical details and fading memories. We will discuss these in various satellite disciplines but not in any order of priority.

## 6.1  Mechanical Systems

Mr A V Patki and Mr H Narayana Murthy came to ISSP from VSSC (then SSTC) to work on spacecraft structures and thermal control systems, respectively, to lay the foundation and both moved to Bangalore in 1972. Dr P S Nair and Dr M S S Prabhu joined structures as design and analysis team leaders and later separated the two activities. Mr S Dasgupta undertook the structure fabrication at HAL Bangalore including technologies like insert bonding to honeycomb panels developed at his group. Mr Samuel joined as a person responsible for testing. They worked in tandem to create capability including facilities and nurturing younger engineers. Dr Sambhasiva Rao later established a Computer Aided Design (CAD) facility to do the design and validation functions concurrently. Dr Nair and Dr Prabhu strengthened structure design and analysis capability to be at par with lead satellite manufacturers, interfacing with launch vehicles, coupled mode analysis, developing test matrix in different configurations, taking up technology development projects and encouraging youngsters to take up higher studies to create human resource. HAL has remained the structure fabrication agency from the beginning and good coordination was provided by Mr S Dasgupta. He nurtured the relationship at the working level and bridged the gap that normally remains between the design agency and the production agency. Structures group interacted with a composite group at VSSC, NAL, etc. for keeping up with upcoming technologies. Dr Nair later became Project Director of INSAT-3B, a fast track satellite to fill the gap due to the loss of INSAT-2D.

The thermal group was developed by Mr H Narayana Murthy with a team that addressed all dimensions. Mr D R Bhandari developed a mathematical model for low Earth satellites and Dr R A Katti developed the same for GEO satellites. Mr V K Kaila coordinated the activities and testing and validation, and Mr Bhojraj developed all needed techniques for implementing the thermal system in satellites. He also took up the indigenisation of most of the thermal materials. Dr A K Sharma developed surface treatment techniques, another important component of thermal systems. Mr P P Gupta and Mr Rustogi developed processes for the VHRR cooler, a tough challenge at that time. Mr Dinesh Kumar developed heat pipes, again a very important component of the thermal control system for high power satellites.

Mechanisms group was a late starter in ISAC as we depended on VSSC in the beginning. It was only after APPLE that Prof Rao realised that ISAC must have a strong mechanisms group and he moved Mr M N Sathyanarayana from integration to lead this newly formed Spacecraft Mechanism Division under mechanical systems. Mr Selvaraj made the solar array deployment mechanism for IRS-1A. Later, Mr M Nageshwara Rao was assigned the development of Boom and Sail assembly and Mr Viswanatha developed a solar panel deployment mechanism for INSAT-2. Mr Sridharamurthy took the development of antenna deployment mechanism. It is to the credit of this early team and to the strict discipline of Mr Sathyanarayana that we have one of the most successful records of mechanism performance in orbit and have never seen any issue with any mechanism developed at ISAC after APPLE.

## 6.2 Power Systems

Aryabhata had Nickel–Cadmium batteries and solar panels from the Soviet Union, but the overall management through power electronics was from ISSP and Mr S Y Ramakrishnan designed the system. Mr R S Mathur started work on solar cells and he was soon joined by Mr Srinivas Murthy who pioneered the solar panel design and fabrication over the next three decades. Surviving over 50 thousand cycles in LEO and the extreme temperatures from −100 deg C to + 100 deg C in GEO requires painstaking efforts for each process parameter, and Mr Srinivasamurthy developed each process through a meticulous testing program. Solar panel fabrication processes for qualification at temperatures of ± 100 deg C proved to be time-consuming but developed very successfully. Mr Kanthimathinathan a BARC trainee joined power electronics to develop an overall spacecraft power system for LEO and GEO missions. Dr Subramanyam, Dr Suresh and Dr Venugopal developed battery assembly and testing and on-board battery management. Dr Suresh gave a very sound methodology of battery design, life cycle management through analysis and testing. Optimum power generation in spinning satellite with continuous changing sun angle for battery tied bus was developed for Aryabhata/Bhaskara satellites and string management for optimal power generation from tracking array was developed for APPLE. Mr B L Agrawal took overall charge of Power Systems in 1977 as Mr Ramakrishnan moved to INSAT-1 project. The large number of charge–discharge cycles with appropriate battery management was developed for IRS-1A. Management of dual power bus, redundancy management and safe mode operations were also developed.

## 6.3   TT&C Systems

Telemetry, Tracking and Command (TT&C Digital) Systems provide connectivity to satellites. Telemetry development was started by Mr D V Raju and Mr R K Rajangam and Ms Rajalakshmi joined to take the activity further. Mr Raju later left for NRSA and Mr Rajangam headed most initial developments. Mr R Ashiya headed Telecommand development and was assisted by Mr D Venkataramana, Mr J P Gupta, Mr U N Das and Mr Y K Singhal. Mr Bhaskaranarayana and Mr U N Das developed the high-power telecommand transmitter. Mr Venkataramana later left to join INMARSAT. ISRO had its own, Telecommand format for about three decades different from prevalent standards. This was the best security feature and the command structure was not much discussed during reviews. The command code for each satellite was kept with only 2 to 3 designers. For operations from an external network, command encoders were supplied by ISRO, working at higher level language. Recently, TT&C has adopted CCSDS standards. Yet another important digital component in Remote Sensing satellites is the on-board memory for data storage and handling it in order to transmit it in a retrievable manner at a much faster rate. It started with 20 MBPS and now has gone up to over 1 gbps. Mr E Vasantha, Ms Veena, Ms Annie Nelson, Ms Vinitha and others were the early contributors.

The TT&C (RF) was initially headed by Mr M K Saha but he left early for INMARSAT. Dr Kosta headed Antenna and major contributors were Dr S Pal, and Mr Lakshmisha and Mr Kista Reddy. Mr Gopal Rao, Mr Kalyanaraman, Mr Sambhasiva Rao, Mr KNS Rao, Mr N Manthaiah and Mr M L N Sastri contributed to RF Systems and Mr Basu developed the Tracking system. Mr Kalyanaraman later moved to the IRS-1A project as DPD and Dr Pal took over as head of RF communication. TTC was in VHF till APPLE, in S band for LEO from IRS-1A onwards and in C band in Geo from INSAT-2A onwards. Mr Lakshmisha also made very unique contributions to Ground Stations by developing High Power feeds. A High Power feed was a very safely guarded technology and Multi-National Companies supply whole ground stations because high power feed was not available in India. Mr Lakshmisha broke that nexus and enabled Indian industry to produce ground stations. Mr Sambhasiva

Rao developed the high data rate transmitters for IRS-1A in X band and the data rates have been going up from a moderate 20 MBPS in IRS-1A to 1 GBPS in Carto-1. Mr K N S Rao made the TTC S band receiver.

## 6.4 AOCS

Attitude and Orbit Control System evolved through each satellite project with the respective development of sensors and actuators. Mr R Sellappan and I joined together in SSD to work on RS-D1 spinning satellite and both moved to Bangalore as part of ISSP. However, Mr Sellappan decided to leave for pursuing Ph.D. at Washington University with Prof Peter M Bainum and I had to take the responsibility from the very beginning. We were developing a magnetic control technique for spin rate and spin axis control for RS-D1 and understanding the space environment from an attitude control point of view like solar radiation pressure, gravity gradient torques, magnetic disturbance torque, radiation belts around the earth, etc. Then the ISSP project came and took precedence. ISSP was a much bigger satellite and Russia had offered a cold gas (pressurised nitrogen)-based spin-up system with six gas bottles, to be operated sequentially, over the life of the satellite. There was no specific requirement of spin axis orientation from scientific payloads. However, we decided to have a nutation damper to damp out nutation motion resulting from thrust misalignment.

The real challenge came from Bhaskara, in which we were to do remote sensing from a spinning satellite. Mr N K Malik had joined as control engineer, Mr S Nagab-hushanam and Mr C B Sathe as mechanical engineers and Mr S Murugesan as elec-tronics engineer to develop control electronics. Mr P S N Rao and Mr Kameswara Rao had come from SSD/VSSC as members of Control Electronics. Mr P N Srinivasan also joined from VSSC as mechanical engineer to head the mechanical engineering activities of the AOCS group. Mr Alok Das, Dr Maharana and Mr S Rajaram joined as control engineers. The Control System Group in ISSP was taking shape. The Bhaskara Spin Axis Orientation and Rate Control System were to be designed with a baseline of a spin-up system of Aryabhata.

I developed a unique technique of 8 pulse logic of spin axis orientation without resulting in nutation using sun vector as reference. Mr Nagabhushanan developed new components like pressure regulator and Mr Murugesan realised the control logic with appropriate telecommand, telemetry and sensor interfaces. To prove that logic works on ground, we had to develop an air-bearing-based angular motion simulator and I got Mr M Nageswara Rao transferred from VSSC. Central Machine Tool Industry (CMTI) machined a 10-inch SS ball and cup assembly with sphericity accuracy of better than 5 microns, a ball to float with an air gap of 25–30 microns. It was a challenging task in those days but the air-bearing simulator did validate our theoretical 8 pulse logic for the spin axis orientation. We developed a magnetic biased control technique to compensate for orbit regression of 4.6 deg per day and this extended the life of the satellite by many years. We also developed automatic control logic to keep the spin axis perpendicular to the orbital plane. Dr T K Alex and his team at the sensors group developed a dual-head Scanning Earth Sensor, an important development, for this purpose. By this time, he had come to Bangalore and was joined by Dr T G K Murthy, a physicist, expert on thin films, and Mr Y K Jain an instrumentation expert. The sensors group worked very closely with the control system group to meet the need for attitude sensors. Mr Kanaka Raju, Mr Koteswara Rao and Mr Kamalakar joined the sensors group and subsequently contributed to the development of attitude sensors of different types.

The next major technology development was the 3-axis momentum biased attitude control system for APPLE. Mr Malik had moved to the INSAT-1 project and Dr P Kudva, a Ph.D. in nonlinear control systems had joined the control systems group. Designing logics for momentum biased system using static Earth Sensor (imported from SAGEM, France) in loop, pitch control by Earth Sensor, roll/yaw control by canted thrusters through roll error threshold cross over, managing spinning phase in the transfer orbit for ABM firing, 3-axis acquisition at 6 am local time, etc. were achieved. One very significant development was new magnetic control logic for roll/yaw control to avoid thruster firing in the on-orbit mode. We have discussed the non-deployment of one solar array story earlier. Mr Ramaswamy joined from CGI/VSSC and took charge of the Hardware in loop simulation activity which he nurtured for more than the next two decades. Mr V K Agrawal joined Mr Murugesan and initiated activity on microprocessors. An 8-bit CMOS processor (RCA 1802) had just been introduced for satellite applications. We decided to use a microprocessor for computationally intensive functions like Kalman filter applications in IRS-1A. A 13-order Kalman filter was designed to model orbital frequency and twice the orbital frequency components in the pitch and roll measurements of the Earth Sensor.

Simultaneously, as lead role for AOCS, we recognised that the development of all elements of AOCS like Earth Sensors, star sensors, Analog Yaw Sensors for yaw update of IRS at poles, dynamically tuned gyros, reaction/momentum wheels, solar array drive control logics like vernier drive, satellite propulsion elements of monopropellants system and unified bipropellant system, mission management, etc. all need very close interaction with control system group. The control system group was to generate specifications, interfaces, combined mathematical modelling for simulations and testing, and responsible for total system engineering. As a result, creating expertise in computer simulation and dedicated computer facility, acquiring state-of-the-art hybrid computer along with a 3-axis servo table became necessary elements of AOCS. Mr Siddhalingaswamy established the computer facility for design, analysis and simulation of end-to-end attitude control system.

We also developed dry lubrication capability with Molybdenum Sulphide through the spray coating route for spacecraft elements. This was developed by Mr Sebastian which became a practice to coat all mechanism components thereafter. We also took the development of ten milli newton electrical thruster for station keeping of geostationary satellites in the mid-eighties. Dr P Prabhakar Rao, a Ph.D. from the United Kingdom initiated this activity. Unfortunately, this development did not come through for almost two decades and the activity was later closed.

The strength of the control group has been in system engineering, putting together all elements, deriving requirements from payload, selecting a set of sensors, actuators like momentum/reaction wheels, magnetic torquers, and propulsion system. It is responsible for designing control logics for all phases of mission and integrating with mission team as well. This includes the ability to simulate from end to end as a total mission.

## 6.5 Integration and Ground Check Out

Satellite Integration is another multi-disciplinary systems engineering activity, largely divided into mechanical and electrical integration. Initially, Mr V A Thomas headed integration with Mr M N Sathyanarayana responsible for mechanical integration and Mr V R Katti, for electrical integration. Mr Pattabhiraman, Mr Dhanabalan and Mr Venkata Rao developed complete mechanical integration support systems starting from integration fixtures, MI, CG measurement facilities, alignment techniques, interfacing techniques using CAD modelling and environment testing at subsystem and spacecraft levels. Mr V R Katti with support from Mr A A Bokil, Mr K S V Seshadri, Mr N Prahlad Rao and Mr M Nageshwara Rao developed electrical integration capability, starting from the basic methodology of documentation, interface matching, EMI analysis, EMI testing and enabling testing of subsystems in various stages of integration of the spacecraft. Dr S V K Sastry, an expert on EMI, developed the EMI laboratory, interference modelling and software tools for interference analysis. Mr Selvaraj and Mr N S Chandrashekar developed Vertical Dynamic Balancing Machine under the overall guidance of Mr V A Thomas. The technology for this unique machine was a closely guarded one with only the US and German firms having it. The machine was extensively used for Aryabhata, Bhaskara satellites. Another machine made for VSSC with remote control facility was used for SLV-3. Perhaps the very first patent of ISAC (ISSP) was taken for this work.

Mr A D Dharma took up the spacecraft Ground Check Out (GCO) activity with Mr O P Sapra, and Mr V R Pratap took up the development of checkout equipment and computer-based testing, respectively, both important for the thorough testing of all elements of spacecraft under different stages of integration. Each project has different requirements of test equipment and payloads in particular and need specialised test equipment. GCO was and is a laborious task, testing all possible combinations in which a spacecraft may be operated. It is generally not allowed to operate a spacecraft in orbit if it has not been tested in that mode in the ground check out. The complexity arises in stimuli generation as a sensor will see in orbit or an actuator responds, and this is a difficult job. A solar array simulator may have to simulate static and dynamic parameters to test the power system performance. Similarly, a battery simulator should simulate battery charge/discharge characteristics for a particular type of battery.

Ground check out generates a designated test sequence so as to cover all operating combinations, measure performance and take a decision on the success or failure of a test. Testing payloads is more complex as the payload is designed outside ISAC, mostly at SAC and hence, the checkout computer should interact with the payload test console, which is designed by the payload team in consultation with the GCO team and QA. Payload checkout system for communication payload for INSAT-2 s was designed by Mr Balasubramayam to totally automate the testing of the payload and completely characterise it.

## 6.6   Spacecraft Mission

Dr A S Prakasa Rao came from PRL to take charge of the mission and soon was joined by Dr Iyengar from TIFR. Mr R S Bhat had the responsibility of orbital mechanics including orbit determination. Mr P J Bhat took responsibility to develop mission software like telecommand processor, telemetry data conversion into engineering units and display and coordination with ISTRAC, a newly formed unit under the leadership of Mr K V Venkatachary with Head Quarters at SHAR. Dr N Ramani established a computer facility to meet the requirement of all technical groups, mainly structures, thermal and mission. Incidentally, Dr Ramani did his Ph.D. in aeronautics but used so many of computers for his thesis that he became a computer expert and remained so, for the rest of his career in ISRO and later in his life. Dr Sundar joined him and played a key role later in the setting up of a big computer centre for INSAT-2.

Mr Y N Bhushan and Mr S K Shiv Kumar joined the mission team after Aryabhata and developed the concept of mission analysis and mission management. Mr Padmanabhan joined Mr R S Bhat for orbital mechanics. Later, Dr M G Chandrasekhar was moved from PPEG to IRS-1A as DPD Mission and as head of mission analysis and mission planning. It included not only satellite on orbit management but also interfaces with ISTRAC (which later moved to Peenya, Bangalore) for LEO satellites, with MCF for GEO missions, post-launch operations and satellite services. Mr Y N Bhushan, Mr Annadurai and Mr Hegde focused on GEO missions while Mr Shiva Kumar and Mr K S Sharma focused on LEO missions. Mr Sethuraman developed the visualisation tool SCHEMACS and was later joined by Mr Eshwara Prakash who developed various software for mission management and data processing needs. The Orbital Mechanics group acquired all the skills and developed software to manage all phases of missions in LEO and GEO.

## 6.7   Technical Physics

Let us recall that Prof Rao was basically a cosmic ray physicist and he along with Dr Kasturirangan had an X-ray Astronomy payload in Aryabhata. Hence, a Technical Physics group was created in the beginning itself with Dr Kasturirangan as head. Mr V Jayaraman, Mr Arun Batra, Y K Jain, Mr Ganage and Dr D P Sharma initiated the development of detectors and electronics for space physics payloads. Mr Y K Jain, Mr K S V Seshadri, Mr Balaram Agrawal and Mr A D Dharma had worked on these activities in PRL. After Aryabhata, some of them got reassigned to different groups and Dr T M K Marar became the new head of Technical Physics.

Mr Jayaraman and Mr Batra moved to SEO core team, Mr Y K Jain to sensors, Mr Seshadri to satellite integration and Mr Balaram Agrawal to head Power Systems. Ms Seetha continued working on astronomy payloads of different kinds. Mr Hatwar provided mechanical design and fabrication support to the Technical Physics group from Aryabhata onwards. Later, Mr Hatwar was moved to MCF Hassan. Many years later, Dr P Sreekumar joined after the retirement of Dr Marar and developed payloads for Astronomy satellites. He was later appointed as director Indian Institute of Astrophysics. This group has many times acted as facilitator to interface between scientists from other institutions like TIFR and PRL and the satellite designers, an important function in developing satellite payloads. The Space Physics laboratory at VSSC has also played a similar role in addition to focusing on some science issues related to atmospheric and planetary physics.

## 6.8   Quality Assurance (QA)

QA has three key elements: (1) Quality Assurance Analysis which acts as an overall mechanism of ultimate quality management, generating QA guide lines for design and testing, conducting Failure Mode Evaluation and Criticality Analysis (FMECA), fault tree analysis, reliability assessment, component-level reliability assessment and specifications, and project level QA plan for achieving the assigned reliability goal through well-structured mechanism to address issues like performance deviations, conformance to design and manufacturing standards. (2) Test and Evaluation (T&E) is another major element which ensures the reliability of any hardware or software through performance and environmental testing from component, subsystem, system and spacecraft levels. (3) Quality Control (QC) refers to ensuring quality of each and every component and subassembly through online checking, inspection and monitoring all processes including fabrication, dimension checks, chemical processes, material testing, etc. Dr T M K Marar was the first head of QA and Mr K R Ramgopal joined him to initiate organising all activities. Mr Siharan De became component expert and all electronics components were procured under his advice. Mr Ramagopal took over QA after Dr Marar became head of Technical Physics. Mr L S Sathyamurthy organised T&E and QC electronics with support from Mr Chockalingam, Mr Nanjudaswamy and Mr H R Nagendra. Mr S G Govindaraghavan organised mechanical aspects of overall QA. Mr V N Purohit and Mr Dave (till his transfer to SAC) looked after Reliability Analysis.

## 6.9   Technical Services

Many allied services for building and testing satellites were pooled together under Dr R D Gambhir, who was primarily responsible for the development of thermo-vacuum facility to test subsystems and the entire satellite. Under him, Mr B N Baliga was assigned to develop and establish environment test facilities like thermo-vacuum chambers, hot and cold chambers and vibration facilities. Mr H R N Katari provided excellent service support as head Engineering Maintenance Division (EMD) and was supported by Mr Rudralingappa. Mr Gopalakrishnan picked from wiring pool of Aryabhata project, set up central electronics fabrication facility under the guidance of Mr V R Katti and continuously upgraded with each project, always keeping electronics/electrical systems production in state of the art. Mr S S Chauhan initially set up mechanical fabrication facility, later Mr Sukumar and Mr L A Jayaraj upgraded it to state-of-the-art prototyping facility as Mr Chauhan left in the early eighties. Dr Mishra joined Mr Baliga to support environmental test facilities.

Environment Test Facilities became a major strength of ISAC enabling tens of subsystems being tested simultaneously. Dozens of hot and cold chambers, thermo-vacuum chambers from half-meter size to 4 m and 9 m Large Space Simulation Chamber (LSSC) and a number of vibration tables in varying capacity were established. LSSC development was constituted as a separate project under INSAT-2. LSSC is a unique facility with 4 m diameter sun beam, a simulator simulating orbital motions all housed in a 9 m thermo-vacuum chamber. INSAT-2A thermal balance test and VHRR cooler tests were conducted successfully in this chamber. Dr Gambhir was the project director of LSSC project with Dr Chandramouli as his deputy. Mr B N Baliga and Mr Srikantaiah provided the technical support.

NAL developed an acoustic test facility in consultation with Dr Gambhir and his team to test a satellite of INSAT class and above. Satellite testing under environmental conditions that a spacecraft may see during launch or in orbit is the key to the success of the satellite program. However, the test program has to be designed very intelligently so as not to be very conservative but also to be effective to prove the reliability under worst-case conditions of operations, including certain non-nominal conditions.

# Chapter 7
# Maturing of the Satellite Centre: The 1990s

This was just the beginning of the journey for the Satellite Centre, setting up all the elements, facilities and putting them together like connecting all the pieces of the jigsaw puzzle. But the real journey of creating a satellite centre that meets all national aspirations was yet to happen. Shri N Pant as director gave a new direction of a disciplined soldier that he was. Though a true follower of Prof U R Rao, he implemented the IRS-1A and INSAT-2A and 2B projects, nurturing the matrix management structure, interfacing with other centres and enabling project directors taking total responsibility in the execution of projects. He set examples of utmost honesty and straightforward management. Once, to pay a visit to a sick friend, he walked to the main road (old airport road) and took a taxi, rather than taking his office car. This was not just for a show, but his real characteristic that he followed strict rules in his own day-to-day behaviour.

As he came from SHAR as director, he had developed a good understanding of launch vehicles and was appointed as Chairman, Mission Readiness Review (MRR), a new mechanism of reviewing overall readiness of all elements of a mission including launch vehicle, satellite, payload, launch preparations and support at SHAR, post-launch support from ISTRAC or MCF. Shri Pant defined how MRR should function and that mechanism continues till date as an effective review mechanism to ensure mission success. His management/interaction with staff association (Joint Consultative Committee, JCM) was also a model of compassion and with frankness. In the year 1990, he was elevated to the post of Deputy Chairman ISRO, possibly to create a place for Dr Kasturirangan to become director of ISAC and later to be elevated as Chairman ISRO.

Dr Kasturirangan's term as Director ISAC saw many successful events as INSAT-2A and 2B got launched in this period; IRS-1C took shape as a real operational class remote sensing satellite, Stretched Rohini Satellite Series (SROSS) for ASLV and IRS-P series of satellites for developmental flight of PSLV. He liked to get into technical discussions with engineers, focused on success, a bit of a conservative, though supportive and seeking more and more analysis. Even as Project Director of IRS-1A, he had nurtured a culture of discussing wider options in open fora, listening

© Indian National Academy of Engineering 2022
P. S. Goel, *Making of a Satellite Centre*,
https://doi.org/10.1007/978-981-16-3480-2_7

to different opinions, and letting the decisions emerge out of discussions. Now, this became a more widely exercised practice, which was slightly in variance with the working of Prof Rao. Professor Rao had extraordinary grasp, asking very relevant questions and assimilating technical aspects very quickly and giving instantaneous decisions. There is nothing wrong with either though Dr Rangan's approach was more practical in a multi-disciplinary environment of a large system, and particularly, if you are not Prof U R Rao.

Professor U R Rao as Chairman ISRO gave full support to Dr Katurirangan and ISAC in terms of growth in projects and infrastructure. They had a special bond, both X-ray physicists from PRL and Prof Rao were responsible for bringing Dr Rangan to ISAC. This served well for the growth of ISAC and Prof Rao could focus on the development of PSLV and other technologies at VSSC including cryogenic engine. He was taking the ISRO agenda forward, perceived by Prof Sarabhai, and nicely articulated by Prof Dhawan with special emphasis on civilian applications of space for the development and growth of the country. ASLV had a successful launch of the SROSS-C satellite on 20 May 1992. However, the PSLV-D1 launch of IRS-1E on 20 September 1993 was not successful. This was a big disappointment to us in ISAC and to the whole ISRO community.

## 7.1 Creation of LEOS

The sensors division headed by Dr Alex and ably assisted by Mr Y K Jain, an instrumentation expert, Dr T G K Murthy, a solid-state physics and thin-film expert and Mr BabuKalakrishnan, a very versatile mind with the ability to address system problems from optics to instrumentation, software, etc. had developed attitude sensors, and the sensors division had created very good strength in spacecraft instrumentation. One of the significant developments by this group under the leadership of Dr Alex was the smart sensor camera for 40 kg RS-D2 satellite using Photo Diode Array detectors, providing a similar resolution as by camera system of the 440 kg Bhaskara.

Based on this strength, Dr Alex proposed the creation of a Unit in the early 90s, on the lines of IISU or APSU, named Laboratory for Electro Optical Systems (LEOS) in the Peenya sheds vacated earlier by ISAC with a bigger mandate encompassing large optics fabrication for optical remote sensing payloads (which were so far imported at huge cost) and other related technologies like thin-film coatings, detector development, etc. Dr Kasturirangan was very supportive and LEOS came into existence as an independent unit, with Director ISAC chairing its management council. The growth of LEOS into a very important laboratory of ISRO and a national-level laboratory in electro optics is largely due to the vision and efforts of Dr Alex, ably supported by Dr T G K Murthy, Mr Y K Jain, Mr Kanakaraju, Mr Koteshwara Rao and Mr C L Nagendra. Fabrication of large optics up to 1.2 m diameter including rough machining, polishing, thin-film coating and light weighting were done for the first time in the country by Dr Murthy and his team at LEOS.

Mr Jain and his team contributed to the development of sensors, particularly Earth Sensors for LEO and GEO missions which are the key sensors for satellites. Mr Koteshwara Rao with Mr G Nagendra later developed star sensor, an emerging technology that had a deep impact on precise satellite attitude sensing independent of the orbit of the satellite. One very unique example of making a QA role as an ideal one was set by Mr H R Nagendra as he got deeply involved himself so much into making sure reliability through design and testing that he had himself developed a deep understanding of these sensors, as good as designers. Normally, the QA role is considered more as a routine, but Mr Nagendra brought a difference to that role. Mr Yalamanchi developed a process of diamond-like coatings, a state-of-the-art technology. Dr C L Nagendra developed the design capability of optical filters through multiple coatings for any desired passband. Dr Alex provided the leadership from conception to a mature electro optics laboratory covering all the disciplines and an assets to the remote sensing program of ISRO in addition to advanced attitude sensors.

## 7.2  Space-Based Services

This period of the early 90s is called a period of transition from an experimental phase of ISRO into an operational phase, largely defined by space applications. In remote sensing, we transitioned to 30 m multispectral resolution from the one km resolution of Bhaskara and the real land use applications started from IRS-1A. We called it an operation satellite era, but an experimental remote sensing era for applications. The National Natural Resources Management System (NNRMS) created by Prof Dhawan, with 13 committees chaired by Secretaries of various departments outside ISRO and developing specific applications using remote sensing data, became very active during this period and Prof Sarabhai's quote "Advanced technology for man and society" started taking shape through this mechanism. NRSA and other Regional Remote Sensing centres started to serve various departments like agriculture, forestry and rural development. Similarly, INSAT-2A and 2B in the early 90s became workhorses to expand telecommunication services, supporting V-SATs in extended C band, TV through S band broadcast and other telecommunication services in C band. We already noted that service requirement was growing much beyond projections and expectations.

I remember a capacity need study of communication transponders by DoT in the late 1980s suggested that INSAT-2A and 2B will be sufficient for the 1990 decade and later we realised that all transponders were fully loaded within a year of launch. Unfortunately, utilization of VHRR payload was only for synoptic assessment of clouds and very little contribution to the weather forecast. This remained a very prestigious payload for IMD with little operational utilisation. Nevertheless, Search and Rescue and Data Relay payloads made very good contributions. Our inability to forecast satellite capacity requirements has caused many blues faced with reality.

It was also time to focus on science missions, particularly to make use of developmental flights of ASLV and PSLV. Perhaps, the first operational class scientific mission was flown on the third developmental flight of PSLV, the IRS-P3 in 1996, in which a TIFR X-ray payload studied designated stars for weeks and months by continuously pointing the satellite telescope towards the stars. It also had a remote sensing payload similar to that flown in IRS-1A requiring earth pointing. We were happy to see a real competition, both User groups asking for more satellite time.

While INSAT-2C and 2D were being realised without VHRR and augmenting the communication transponders, INTELSAT, a major player as an international satellite communication provider, became interested in leasing satellite transponders from an Indian satellite at Indian slot by leasing the capacity and on some frontend payment. Mr Narayanan director SATCOM office at ISRO Head Quarters and Mr P Ramachandran were pursuing very hard to make this agreement happen. The agreement with INTELSAT was signed in January 1995 by Chairman, ISRO, Dr Kasturirangan, Mr K Narayanan and Mr V R Katti. The payload beam was to be shaped to cover landmass from central Europe to Australia. The payload was complex with a dual grid-shaped beam for landmass coverage in dual polarisation.

We had no capability to develop a shaped beam dual gridded antenna in ISRO hence, decided to import the antenna. There were two major gains for ISRO from this agreement, namely (i) To enter into real operational, state-of-the-art communication satellite technology of the time and (ii) To get a kind of international recognition that we had indeed achieved the operational class communication satellite building capability. We had to build test facilities like "Compact Antenna Test Range" at ISAC and SAC, though the facilities got ready a bit later (after the launch of INSAT-2E) and as such we got the antenna tested at the vendor's end.

Meanwhile, a major transformation was taking place in TV broadcasting. INSAT-1B had created a large network of cable operators receiving feed in S band and retransmitting through cables to reach individual homes. This was competing with Doordarshan's network of re-diffusion, particularly in cities and the cable operators were adding additional channels from other international satellites as well. A new business model had emerged. But the S band receive antenna was too big, about 11 m and hence expensive. It was decided to simultaneously transmit the content in C band also, at relatively lower overhead due to the availability of larger bandwidth in this band. This reduced the antenna size at the cable operator's end to 7.5 m, substantially reducing the operator's cost. Ground antenna system costs cube (power 3) of the antenna size. Slowly but steadily, the complete network shifted to C band by the late 90s, releasing the S band broadcasting bandwidth for other services.

## 7.3   The Journey in Mid-90s

We may say that the setting up of ISAC as a Satellite Centre was complete in the mid-90s with all basic skills of designing, developing, fabricating and testing of all types of satellites; remote sensing, communication and space sciences established.

Simultaneously, SAC had matured as a professional payload development agency for remote sensing and communication payloads in addition to its other important role as developing space applications in all areas including meteorology and atmospheric sciences. ISAC had developed unique management culture to interact with sister agencies and got satellite subsystems in the right quality from SAC, LPSC, IISU, composite group of VSSC and external agencies like HAL, NAL including some private companies. Even though the products came from different agencies, the complete system engineering remained with ISAC. This is important as the lead centre for satellite systems. Once again, credit should go to the leadership of Prof U R Rao. He often used to say that the overall responsibility lies with the corresponding group or project team at ISAC. However, the journey of a development agency like a satellite centre would never be complete. It is a continuously evolving process with new challenges emerging every new day. *That is why, on the one hand, I say that the making of the satellite centre was complete by mid-90s, on the other hand, the process continues even today. We had developed the capability, but not the capacity. We were making only one to two satellites in a year.*

Professor U R Rao retired as chairman, ISRO, in end of March 1994 and Dr Kasturirangan took over as new Chairman of ISRO. Mr R Aravamudan was appointed as the fourth Director of ISAC. An expert on Radars, Mr Aravamudan was appointed in ISRO by the founder Prof Sarabhai himself. He was director of SHAR centre and had wide exposure to ISRO activities. After the departure of Dr Kosta, Dr S Pal had taken overall responsibility for RF Communication System. Mr Tarsem Singh retired on superannuation. However, Dr Prakash Rao, Mr Patki, Mr Thomas, Mr Sathyanarayana and most of the members of the Aryabhata team were still available. Mr Ashiya was back as Advisor to Director ISAC from his assignment as Associate Project Director and Resident Director, INSAT-1C/D.

Professor Rao had initiated the concept of Technology Development Projects (TDP) to meet future requirements as technology was changing fast and we had to be on par with state-of-the-art technologies. VSSC had emerged as lead centre for launch vehicles with LPSC and SHAR as associate centres. Also, IISU and Composite unit (REPLACE) at Trivandrum were associate units, though these units had larger workload and funding from satellite programs. ISAC had emerged as the lead centre for satellites. LEOS became the associate unit of ISAC providing the sensors to the satellites and optics to imaging payloads. SAC focused on space applications as the lead centre with additional responsibility to provide applications payloads for satellites. NRSA, though an autonomous unit under DoS, acted as an associate unit of ISAC for satellite mission planning and satellite operations planning for remote sensing.

MCF, Hassan started to grow as the number of communication satellites were growing and all its growth was largely funded through INSAT projects from ISAC. To provide space diversity to satellite operations, it was decided to have one more location of MCF at Bhopal. A very harmonious relationship was fostered between MCF operational team and ISAC/INSAT mission team. Similarly, ISTRAC also needed continuous expansion and support from the IRS mission team at ISAC for smooth operations. Both MCF and ISTRAC have developed as associate units. This is the most important aspect of evolving culture of ISRO that so many independent units can work so harmoniously for national missions.

LEOS developed itself into a lead unit in large optics fabrication including related processes like polishing and coatings in addition to attitude sensors. NRSA was responsible for data archival from remote sensing satellites and dissemination to all users in the government and industry. Applications developed by SAC for remote sensing data were to be transferred to NRSA and Regional Remote Sensing Centres for wider usage of the satellite remote sensing data. Theoretically, NRSA, an autonomous body under DoS, was not part of ISRO, but it was having smooth coordination with the ISAC mission team and SAC data products group.

Our main programs at ISAC were (i) INSAT-2C, 2D and 2E for communication, (ii) IRS-1C,1D for remote sensing, and (iii) building payloads (satellites) for developmental flights of launch vehicles, the IRS P series (P for PSLV) and SROSS series (for ASLV), some accommodating science missions. The most challenging missions with new technologies were IRS-P3, IRS-1C, IRS-P4 and INSAT-2E.

## 7.4   IRS-P3

IRS-P3 was being developed for the third development flight of PSLV and initially, it was decided to have a low-cost remote sensing payload, a multispectral camera, the payload for the developmental launch and augment the multispectral coverage. Professor P C Agrawal from TIFR came up with a requirement of accommodating an X-ray astronomy payload with the accurate pointing of payload axis towards any designated star for weeks and months and changing from one star to another. This required inertial pointing capability by reaction wheels from any orientation to any other orientation that we had not developed yet. The DTGs in IRS-1A/B had given confidence and we were developing a 1750 microprocessor-based Attitude Control Electronics, hence we saw this as an opportunity. Mr K Thyagarajan was Project Director for IRS-P3 and was supportive of this venture. Mr N Venkateswaran had been studying the quaternion referencing scheme and he developed the algorithms. Once the star was in the field of view, we could derive reference from the payload, though in a very narrow field of view. IRS-P3 became a multi-mission satellite. This experimental satellite became so popular with remote sensing data users as well as with TIFR scientists that there was always demand from both the groups for satellite time. The satellite originally designed for one-year operational life worked satisfactorily for almost a decade.

## 7.5   IRS-1C/D (Resource Sat-1 and 2)

After the unqualified success of IRS-1A and 1B, we were looking for a landmark remote sensing satellite and decided to build a next generation remote sensing satellite as the state-of-the-art satellite. Mr S Kalyanaraman was appointed as Project Director. At that time, the LANDSAT-4 satellite being developed by NASA at a cost of about 1.2 billion dollars was making news as a great development. We decided to match the payload capability in bands and improve in resolution. A 23 m resolution multispectral camera in four bands, a 5.8 m resolution panchromatic camera and a 70 m resolution wide-field vegetation index monitor with 2-day repeativity were selected as payloads. It required a totally new configuration, a high data rate of 125 mbps and a 12-channel tape recorder to collect data from anywhere in the world. IRS-1C was launched on 28 December 1995 by Russian launch vehicle Molinya from Baikanur and its twin IRS-1D was launched by PSLV-C1, the first operational flight, on 29 September 1997.

IRS-1C remains the highest point in Indian remote sensing satellite history that gained overnight recognition as the best remote sensing satellite in the world. Once in 1998, the Chief of Russian Space Agency came to visit us in ISAC and having seen data from IRS-1C and IRS-P4 in the cleanroom, he remarked, "ISRO is about 10 years ahead of Russia in remote sensing satellites". It was so pleasing to listen to this comment from the Chief of Russian Space Agency. We (Dr George Joseph, Dr Ravi Dutt Sharma, then scientific secretary, and myself) had visited Moscow in the early eighties to understand all about satellite remote sensing in the USSR. They did not show us any hardware and just explained general methodologies. After 15 years, the scene had reversed.

## 7.6   IRS-P4/ Oceansat 1

On a demand from Ocean Resource Community in 1996, it was decided to make a special satellite for studies of Oceans on an operational flight of PSLV. By now, we had two successful flights of PSLV and could go for an operational satellite launch. An 8-channel Ocean Colour Monitor with high sensitivity and a Scanning Radiometer with high gain antenna were selected as the state-of-the-art payloads. Mr R N Tyagi was Project Director. The observation time was selected as 12.00 noon to maximise the signal as ocean reflectivity is rather low. Huge scan mechanism development and its accommodation was a big challenge. Once again, the payload and satellite bus performed well setting a new bar. Figure 7.1 shows IRS-P4 in cleanroom. The Potential Fishing Zone (PFZ) application developed at SAC by the group led by Dr Shailesh Nayak using IRS-P4 data has made a huge difference to the fishing community of India, saving thousands of crores worth diesel every year. Today, about six lakh fishermen are taking advantage of PFZ advisories issued by Indian National Centre for Ocean Information Services, Mistry of Earth Science.

**Fig. 7.1** IRS-P4

IRS-P4 undergoing dynamic balancing test (1998-99)

The Geostationary Satellite Launch Vehicle (GSLV) development started in the early nineties to launch a 2000 kg communication satellite into 170–36,000 km, 19 deg inclined Geo Transfer Orbit (GTO), and this required development of a new satellite bus. After INSAT-2E, we had the sanction of INSAT-3 series satellites with higher payload and 10 years plus life. This had to be in three-ton weight class, considering the dual launch capability of the Ariane launch vehicle. We called these I2K and I3K buses, respectively. Both had similarities in avionics, architecture but were different in size, in terms of propellant loading, thermal dissipation and power generation. The payload capability for I3K was almost twice that of I2K.

By the mid-nineties, ISAC was organised into four major technical areas and a support group: (i) Mechanical Systems Area headed by DD, Mr Patki, (ii) Electronics Systems Area headed by DD, Mr Ramachandran, (iii) Control and Mission Area headed by me as DD, (iv) Integration and Ground Checkout Area headed by DD, Dr Prakasa Rao. The facility group consisting of Mechanical Workshop, Electronics Fabrication Facility, Environmental Test Facilities and related services was headed by General Manager, Mr Baliga. In addition, the Geo Satellite Program was headed by Mr V R Katti and Remote Sensing Program was headed by Mr Kalyanaraman, Small satellite Program was headed by Mr Thyagarajan, Reliability and Quality Assurance Group was headed by Mr M N Sathyanarayana and PPEG by Mr V A Thomas.

## 7.7 Maturing Satellite Technologies

### AOCS

Our strength in Control Systems grew consistently, along with the Centre to about 80 persons including AOCE, Magnetic Torquers and related mechanical hardware, Electrical Propulsion, Dry Lubrication, and Attitude Control including Dynamics Modelling, Navigation, Attitude Sensor interface and complete AOCS system engineering including simulations. As a system engineering group, it was the whole of Attitude and Orbit Control System (AOCS) that was my responsibility meaning an anchor's role, with IISU for the development of inertial sensors like gyros, reaction/momentum wheels and solar array drive, with LPSC, Bangalore unit on the development of Mono Propulsion System for LEO satellites, Unified Bipropellant System for GEO satellites and qualification of its components, with LEOS for development of attitude sensors like Earth Sensors and star sensors and with Flight Dynamics group in ISAC for orbit maintenance strategy.

Within the group, the challenge was always to prepare for the future. We focused on study and modelling of flexibility and multibody dynamics, spot imaging with agile satellite as an example of IKONOS, highly stable satellite for low jitter, precision arc minute pointing capability anywhere to anywhere in space, re-entry and rescue capability, docking in space, electrical propulsion, magnetically suspended reaction/momentum wheels, alternate propulsion systems like superheated water, attitude and orbit control-related problems of very large interconnected bodies like space station and solar power satellite in GEO with dimensions spreading up to 11 kms, low earth constellations for communication and formation flying for radar interferometry, etc. With a priority on deliverables to projects, some of these activities remained in studies only.

One fact we realised was that the best way of developing technology capability was to make it a part of our forthcoming projects and taking the development of technology within the project schedule. This may sound a totally unaccepted philosophy in today's context, but it worked very well. There are some technologies that we could not succeed in spite of at least one dedicated engineer. Electrical propulsion and magnetic suspension are two such examples. Initial progress was encouraging. We did develop a prototype of a 10 milli Newton Argon thruster, a one metre test chamber with thrust measurement capability (to an accuracy of 1 milli newton) and related instrumentation, but proving reliable operation for station keeping remained elusive. Development of cathode with a life of 10,000 h could not be achieved or procured. LPSC took the development some two decades later; we now have an electrical thruster in ISRO.

Magnetically suspended wheel development got interrupted by the concerned engineer (Mr Rajaram) going abroad and the development was later undertaken by IISU. We established a very good Hybrid Computer facility during APPLE and used it effectively till IRS-1A and 1B, but later moved to all digital simulations during INSAT-2, adding a 3-axis servo table for hardware in loop simulations for IRS-1 and INSAT-2. Adding a separate computer facility, in addition to a central computing facility at ISAC, was necessitated as simulation and on-board system emulation was a specific requirement of AOCS and it continues to be so.

During Bhaskara days, with heavy commitment to developing cold gas propulsion systems including components like flow control valves and pressure regulators, we even had established our own mechanical fabrication facility in the control system group, which was later merged with the central fabrication facility. Mr Nagabhushanam played a key role in these developments, including electrical propulsion. He headed the mechanical activity in AOCS for a long time. Mr Sebastian developed dry lubrication technology for spacecraft mechanisms and the technology was transferred to industry.

Dr V K Agrawal took lead to set up control electronics activity including state-of-the-art digital laboratory to develop Attitude and Orbit Control Electronics which incorporated evolving microprocessors: starting from RCA 1802 (8-bit processor) implementing 13-order Kalman filter to model first- and second-order orbit components in Earth Sensors in IRS-1A and 1B, 16-bit 80C86 INTEL microprocessor in INSAT-2A/2B onwards as total digital controller, Mil 1750 microprocessor from INSAT-2E onwards, introducing custom-designed ASICs and FPGAs, increasing functionality at reduced mass and power and ultimately developing a multifunction Bus Management Unit (BMU) incorporating AOCS, TM and TC digital functions in the early 2000s. It was a very significant contribution to the ISAC journey of evolving technology towards small satellites.

After Dr Agrawal left, the activity has been further continued by Mr Udupa and Mr Sudhakar. We had imposed certain software disciplines onto ourselves, over and above routine software practices for critical systems of software engineering, documentation and testing, like the use of a well-defined safe set of instructions, use of very well-tested compilers even at the cost of obsolescence, non-usage of interrupts of any kind to make software/firmware 100% testable, not exercising any option in orbit that has not been fully tested on ground, etc. It is due to this discipline that we have not had any on-orbit bug or software anomaly in any AOCS in its history over three decades.

Mr Malik returned to AOCS after his assignment in INSAT-1, just in time to take a lead role in designing AOCS for INSAT-2A. Dr Kudva left for the US for a teaching assignment. With his exposure to INSAT-1 AOCS design for a stable 3-axis stabilised design with momentum management in V-mode (two canted momentum wheels) and L-mode (one canted momentum wheel and one lateral reaction wheel), Mr Malik took overall responsibility with a few youngsters (Mr Kulkarni, Mr Siva, Mr Jaswinder Singh Khoral, etc.) and very successfully developed the system with very effective coordination with the mission team. All INSAT-2, 3 and 4 satellites have retained this design feature for an efficient AOCS in terms of meeting requirements of various complex missions and efficient mission management from MCF.

The highlight of the robustness of the design came from INSAT-2E in which both the Earth Sensors failed early and we had no substitute for an international commitment of INTELSAT. Mr Malik developed a strategy of open-loop attitude control of roll and yaw axis using momentum history and using gyro for pitch axis. He derived attitude measurement from infrequent measurements from VHRR images and provided momentum bias as per the previous day's history. The technique worked for the complete mission life of INSAT 2E without the user noticing the failure of Earth Sensors. This is perhaps unprecedented in the history of space, though not much talked about.

Mr N Venkateswaran developed an inertial reference system taking incremental angles from DTG using quaternion, and we implemented a capability to move a satellite from any orientation to any other orientation for IRS-P3. Mr Venkateswarlu and Mr Natarajan focused on the design of control systems for Remote Sensing satellites and Mr Natarajan continued, after Mr Venkateswaralu left for a private company. Later, many new engineers along with Mr N K Philip joined and the legacy continues continuously evolving the design for better stability, lower jitter, high agility for ever-evolving higher resolution in remote sensing satellites. One important milestone of adaptation of zero momentum-based design is the IRNSS satellites in which satellite north–south has to be flipped once in 6 months for meeting the power requirements and the operations are routinely done. Still, the ultimate achievement is in IRS-3B series of satellites with very high resolution (0.28 m PAN and 1.13 m multispectral), very heavy satellite of 1600 kg and still agile for spot imaging. It is heartening to note that ISRO is able to retain the leadership role in remote sensing satellites for more than two decades, which started with IRS-1C in 1996. This is a multi-disciplinary achievement enabled by the payload team at SAC, AOCS and communication teams at ISAC, control moment actuators at IISU and the whole system engineering efforts of AIT and Mission.

**Structures**

Taking new technologies as part of the project had been contributing to the technology development in all areas and facilities. Dr P S Nair and Dr M S S Prabhu duo under the overall guidance of Mr A V Patki developed a total capability for design, analysis and testing of satellite structures with Dr Nair specially focusing on geostationary missions and Dr Prabhu on LEO missions. Mr S Dasgupta was interfacing with HAL as space structures fabrication centre. Mr Samuel established a state-of-the-art structure testing facility to validate the design of any new structure with heavy instrumentation capability. The group developed the capability to model structures with state-of-the-art software like NASTRAN with the capability to perform thermal analysis of packages. Also, expertise was developed to generate reduced models for coupled mode analysis with the launch vehicle and also run the coupled mode if the launch vehicle provides the equivalent model. Dr Sambhasiva Rao set up a separate laboratory for Computer Aided Design, concurrent analysis and generated machine-compatible drawings/code for manufacturing on a numerical machine. Dr Prabhu later left ISAC to take a senior-level position in TCS and later moved to Infosys.

The structures group had clarity from the very beginning that fabrication of space-craft structures will be undertaken at HAL and HAL also committed itself by creating a separate division, establishing all facilities exclusively for satellite structures, in particular the composite structures with aluminium and fibre glass. Many experimental studies like insert bonding or mechanical packaging were done using in-house facilities and were transferred to HAL for production. Many studies have been conducted to compare the efficiency of spacecraft structures as a percentage of overall spacecraft mass. There is a systemic reduction in each category of satellites and its efficiency matches well with contemporary designs elsewhere; however, there is a considerable need for R&D in this ever-evolving field of new materials and newer designs. Larger adaptation of carbon composite structures, smart structures with embedded sensors and adaptive structures with varying operating environments like thermal have been the focus of R&D. Dr Renji and Dr Srinivasan have provided a leadership role in subsequent years.

We initiated a study on a lighter than air stratospheric platform with Dr P S Nair in lead. These have huge potential as pseudo satellites for local communication.

**Thermal**

Thermal control in satellites is very critical for the whole mission life and requires a deep understanding of the space environment including radiations coming from the Sun, Moon and Earth, spacecraft geometry, the ability to create thermal model at the overall spacecraft level and also at subsystems package level and sometimes even at PCB/component level. It is important to adequately characterise all thermal materials for their thermal properties and their degradation, implementation of design with perfection and evaluate overall performance through thermo-vacuum and thermal balance testing.

Evolution of the technology took place under the guidance of Mr H Narayana Murthy. Mr D R Bhandari developed thermal modelling capability for spinning as well as for 3-axis stabilisation in LEO. LEO satellites have an orbital period of about 100 min including a typical one-third of the time in eclipse, hence, averaging takes place over 100 min. APPLE was the first GEO satellite and hence 24 h orbital period for 3-axis stabilised satellite means the same face facing the sun for a long period, increasing the complexity of the thermal design. Dr R A Katti with the guidance of Mr H Narayanamurthy made the thermal model for APPLE and did the design which adequately worked in both spinning and 3-axis stabilised conditions even with one side solar panel not deployed. He along with Mr V K Kaila took the lead and worked on thermal design and analysis for the INSAT-2 series of projects pooling the expertise gained in APPLE and INSAT-1 projects.

While a special thermal control validation test in thermo-vacuum chamber was adequate for APPLE since it was a low-power satellite, a thermal balance test was necessary for INSAT-2. Thermal balance test conducted for INSAT-2A in LSSC proved the adequacy and robustness of thermal design and also proved satisfactory prediction of thermal regimes based on thermal modelling. The thermal control system of INSAT-2A was satisfactory and the capability to design and implement the thermal design for GEO missions was established. The person behind the implementation of thermal design was Mr Bhojaraj. He was well supported by Mr P P Gupta, Mr Rustogi and Mr Dinesh Kumar. Mr Rustogi played a key role in this development cryogenic cooler for cooling the infrared detector of VHRR to about 100 K. We had a failure of thermocouples in VHRR cooler of INSAT-2A and 2B, a few years after launch, rendering VHRR unusable. Mr Rustogi took a painstaking testing exercise in a thermo-vacuum chamber for a year to understand soldering/bonding problems for a joint that undergoes temperature cycling at low temperatures.

Mr Dinesh Kumar developed the heat pipes for usage in high power communication satellites to transfer heat from a source like TWTs to radiating surfaces on north and south panels. Production of heat pipes in the industry on technology transfer has been a major achievement. Mr Bhojaraj undertook the development of thermal materials like MLI and OSR in the industry as these were being imported and there were only one or two sources in the world. Development of active sterling cryo-cooler was also taken up with IIT, Mumbai, though with limited success and actual qualification came two decades later. The thermal group was ready to undertake any mission from design, mathematical modelling and implementation to testing at component, subsystem and spacecraft levels. The thermal group very successfully designed Chandrayaan and Mangalyaan thermal control systems, which is the ultimate test of the capability of the group.

**Spacecraft Mechanism**

One area that continues and will continue to demand new developments perhaps for many years to come is spacecraft mechanisms. Some areas like electronics are driven by civil technology developments, and we adopt in space with a time lag of 5–10 years allowing the technology to mature, but Spacecraft Mechanisms are unique to space. Starting from Solar Panel Deployment Mechanism of IRS-1A, development of sail and boom assembly for INSAT-2A, Antenna deployment mechanism unique to each spacecraft in multiple numbers, detaching system for scan mechanism for Oceansat-1, deployable antenna, rotary joints for scanning antennae, etc. all have been developed and flown successfully. Mr M N Sathyanarayana was a leader with perfection in engineering and he nurtured youngsters with strict discipline and emphasis on testing the mechanism under worst conditions to establish design margins. Mr Nageshwara Rao joined him after completing the air-bearing simulation system at control system and Mr N C Bhat joined from the structures group. Mr B S Nataraju, Mr C D Sridhara, Mr Sridhara Murthy and Mr Viswanatha were part of the young team that developed into a leadership role.

The future communication satellites with high throughput capability will require 12, 18 and 23 m unfurlable antennae with many feeds and deployment of large solar panels generating 20–25 KW of power. The mission to Moon or Mars will require many complicated mechanisms each on Lander and Rover. The same is true for interplanetary missions. Ultimately, we are looking for a humanoid robot doing complicated functions on Moon or Mars before man could land for a long duration mission.

**Power Systems**

The Power group has developed the capability to design complete spacecraft power management, including battery charging, eclipse management, battery sizing for different missions, array power generation management, etc. Mr Kanthimathinathan, Mr S T Venkataramana and Mr Ranganath Ekkundi were all part of Mr Balaram Agrawal's team and helped to develop systems for LEO and GEO missions. Post

Aryabhata, Mr Mathur left for ITLU Paris and Mr Srinivasa Murthy took over the solar panel activity, qualifying critical processes like solar cell interconnect welding, fabricating panels with packing density commensurate with the world standards. Expertise has been developed to design with various kinds of solar cells, including multi-junction GaAs solar cells. Dr M R Suresh developed expertise in spacecraft batteries. While Ni-Cd batteries remained the core of LEO missions, Ni-H2 batteries and later on Li-ion batteries were adopted for GEO missions for their higher energy density. This group along with Dr Venugopal developed the ability to design an appropriate battery for a given mission and then testing for adequacy of the design. All test and storage facilities were developed in-house.

INSAT-2A was a major technology upgradation exercise to make the power system compact and reliable with mounting of shunt box outside the spacecraft, on the yoke. Mr S T Venkataramanan took lead in designing a compact modular power system for GEO satellites. The concept of end-to-end redundancy management had emerged and the space grounding scheme had been evolved with the spacecraft Integration group taking the lead. The dual power bus management was evolved with shunt regulator for GEO mission accommodated outside the spacecraft.

Power systems undertook a major exercise of developing HMCs for DC–DC converters and transfer technology to the industry as HMCs reduce weight, improve reliability and increase productivity. Mr Sankaran joined the solar panel group and played a key role in standardising panel fabrication processes. He later headed the group and developed overall design and testing capability up to 12 KW in GEO.

### TTC and Data Handling

Telemetry, Tracking and (Tele)Command are very basic to satellite technology to provide access to the satellite. Aryabhata had a very simple VHF system with only 20 commands and 256 bit per second telemetry. These were realised with basic SSI CMOS gates available in the early 70s. Tracking support came from ranging tones via the TC receiver and TM transmitter. Though seen together functionally, TC, TM and Data Handling have been with the digital group and Tracking has been with the RF group along with receiver and transmitter. A continuous evolution has gone through with each series of satellites in increasing order of complexity.

The payload data in Bhaskara was at 100 kbps in a separate carrier with increased transmitter power. The concept of Data handling really came in IRS-1A with 20 MBPS payload data in the S band which was later increased to 85 MBPS in IRS-1C. Normally, it was preferred to have TTC in payload band from a frequency coordination point of view in a communication satellite, though not with very obvious engineering advantage, hence the C band was adopted for the INSAT-2 series of satellites. By the mid-nineties, we were standardising at 1 kbps data rate for TC, 1 kbps data rate TM and data rate of 150 mbps in X band for remote sensing satellites. A real challenge came in Radar Imaging Satellite (RI SAT) with a data rate of 640 mbps. As payload data rate is an ever-increasing requirement, near loss less data compression techniques became inevitable, in spite of the reluctance of data processing and user groups. It was first adopted in IRS-1C with a high-density tape recorder procured from the US. Key persons behind these developments were Mr R

K Rajangam, Mr J P Gupta, Mr Y K Singhal and Mr Lakshminarasimhan for TC, Mr U N Das, Mr R Seshaiah and Mrs Rajalakshmi for telemetry and data compression, Mrs Annie Nelson and Mrs Vanitha for Data Handling, and Mr Vasantha and Mrs Veena for solid-state recorders.

**RF Communication**

Like TTC, RF has also evolved through projects to cater to the need of mainframe and payload data communication, starting from the VHF low-power transmitter and receiver developed by Mr Saha, Mr Kalyanaraman, Mr Gopala Rao, Mr M L N Sastry and Mr Manthaiah. We still had a VHF band for APPLE for Geo. Major development came for IRS-1 for TTC in the S band and data transmitter in X band. C band TTC was developed for INSAT-2 satellites. Dr S Pal led the team for S, C and X band RF TTC. The on-board RF system consisting of on-board Telecommand receiver, telemetry transmitter, transponder mode for ranging and antennae. The key persons who made significant contributions to these developments include Mr P Kista Reddy, Mr K N S Rao, Mr V Sambasiva Rao, Mr Mahadevan and others. Mr Lakshmisha played a key role in developing omni-directional antenna, a unique requirement for satellites. One of the pioneering concepts of developing range compensating data transmitting antenna for IRS satellites by this group was an innovative idea before the advent of a dual gimballed tracking antenna. The tracking system development has been the forte of Mr S G Basu from the very beginning. He developed a robust interfacing mechanism with the mission team to process the data for orbit determination. Later in the mid-nineties, he realised that the GPS receiver is going to be an important element of accurate orbit determination of LEO satellites; he focused on GPS receiver development in the industry. Communication Systems group also developed a unique spherical Phased Array antenna in X band for TES satellite which could generate two beams simultaneously enabling the transfer of data to two differently located ground stations.

**Satellite Integration**

Satellite integration is much more than putting two or more subsystems together. It is an exercise of complete system engineering accounting for payloads field of view, interfacing with all subsystems for physical accommodation with their individual constraints, minimising harness, looking into radiation effect on vulnerable components, thermal environment of each component and ease of integration and correction on the ground. It is a concurrent exercise and sometimes may take five to six iterations before an acceptable design emerges. This discipline was first headed by Mr V A Thomas. He, being a mechanical engineer, established the mechanical integration methodology, created infrastructure like developing jigs and fixtures for mechanical integration and set up facilities for CG, MI and dynamic unbalance measurements.

Mr V R Katti supported him for electrical integration, setting standards for interfaces, creating electrical integration setup, laying harness, etc. He also set up an EMI facility for testing subsystem susceptibility and certification. Grounding is one of the important considerations in satellite integration, and charge accumulation on any small element can result in disastrous consequences. Mr Katti along with his colleagues Mr K S V Seshadri, Mr N Prahlad Rao and Mr A A Bokil conducted numerous experimental studies to come up with various schemes for ISRO satellites. Selection of wire for electrical harness, testing of harness to survive in LEO and GEO missions and joining a wire in harness, each needed to be standardised. Integration of a satellite is an iterative exercise through electrical and 3D mechanical modelling between payload, sensors, subsystems, battery and thermal control. ISRO has created this expertise not only for standard remote sensing and communication satellites but also for complex scientific interplanetary missions. Another area in which intensive efforts put in by this is intra-system and inter-system Electromagnetic Compatibility. Dr S V K Shastry, M Nageshwararao for theoretical modelling and EMC Software development, Mr T Parthasarathy, Mr K N Shamanna ( both joined ADA later) and Mr C S Nagaraj (left for Wipro) for establishing EMI test facilities and Dr Hariharan for Electro Static Discharge (ESD) analysis and testing, together formed one of the strongest groups on EMI/EMC and ESD in the country.

**Ground Check Out**

Ground Check Out of a satellite is an elaborate testing exercise in different phases of satellite integration. Ground check out involves powering the spacecraft through battery simulator and solar array simulator, generating input stimuli for sensors testing, check out of payload by creating similar inputs like the satellite will receive in orbit, data processing of payload to assess its performance and generating all possible combinations of satellite operations for mainframe and payload. Mr Tarsem Singh started this activity with Mr V R Pratap developing a computer checkout system and Mr O P Sapra developing hardware for powering and stimuli generation for sensors and other systems during the Aryabhata project. Post Aryabhata, Mr A D Dharma took over the reins from Mr Tarsem Singh. This combination worked together for almost three decades and developed the state-of-the-art checkout systems, improving with every new project. Initially, the payload checkout equipment used to come along with the payload, but later the group developed expertise to design and develop their test equipment, except for a very specialised payload. The communication payload checkout system was totally developed by Mr Balasubramanian who came from SAC and Mr C R Srinivasan transferred from INSAT.

One important guideline we adopted was that any mode of operating a satellite in orbit would be taken up only if it had been tested a priori in the ground checkout mode. This forces working out a very comprehensive test plan during the AIT phase before launch.

**Quality Assurance**

Quality Assurance group is the largest group in ISAC catering to QA at different levels. At the very top level is the Quality Analysis providing analytical support to quality. This consists of reliability apportionment to various components of the mission, FMECA analysis of each subsystem and overall system, review of quality system through configuration control board, waiver review board, System Review Board and Sub System Review boards for each subsystem. On top them all, project independent Standing Review Committee are created as per the comprehensive product assurance plan for a given project. At the next level is the Test and Evaluation of each subsystem through creating appropriate test matrices and documenting even minor deviations. The third level is Quality Control (QC) in which quality is ensured through online inspection including incoming parts and material control and each process being certified for its stability, reliability and space worthiness.

Mr Siharan De, the component expert, laid a strong foundation for the space-qualified parts/material procurement activity, standardised parts screening method-ologies and worked on techniques for upgrading and qualifying components for space use. Mr L S Sathyamurthy played a key role in evolving test and evaluation proce-dures at subsystem level and standardisation of thermo-vacuum tests at the spacecraft level. Mr S G Govind Raghavan worked on tools and methods for the validation of mechanical systems inclusive of their qualification and acceptance. Mr V N Purohit initiated and nurtured work on reliability analysis along with Mr Dave who even-tually moved over to the SAC remote sensing payload group and was responsible for the design of VHRR electronics. Mr T S Nanjudaswamy developed expertise on quality control of electronics fabrication processes and organised a large group for QC. Most of the guidelines for QC of mechanical components and electronics fabri-cation were evolved through detailed experimentation. Mr K R Ramgopal continued to be head of QA till he opted for VRS to start his own enterprise. Mr Gopalkrishnan, in charge of CEFF, played a key role in the development of electronics fabrication and packaging techniques with built-in QA procedures.

**Process Qualification**

Development of indigenous technology is really no rocket science but mastering various processes that go into the manufacturing of satellite subsystems and docu-menting these is the real challenge. Getting into minute details of a process, control-ling its viability through a formal document, introducing checkpoints for inspection, maintaining traceability of subprocesses, materials and equipment used, recording the calibration of equipment used, etc. are needed. This applies to all disciplines and a simple activity like hybrid micro-packaging may involve 30–40 processes. In other words, the ability to control processes for high repeatability that will not allow a defect to creep in is actually quite formidable. In our effort to indigenise a large number of subsystems from MLI, coatings, solar panel, etc. we needed to write elab-orate details and qualify the process. Mr T S Nanjundaswamy from QA took that task on himself. Many of these might be an adaptation from the available literature, but a lot of thinking with respect to practical limitations of equipment or skill is required. Mr Nanjundaswamy's role in defining the process and its qualification was very important in bringing industry into taking satellite activity and indigenisation.

**Environment Test Facilities**

With a modest start of a few hot and cold chambers, a 4-ton vibration shaker and a one-metre thermo-vacuum chamber, the team of Dr R D Gambhir and Mr Baliga continuously upgraded the environmental test facilities during the APPLE project by adding almost a dozen of hot and cold chambers, a 2 m thermo-vacuum chamber. A major upgradation of ETF took place under the INSAT-2 project through realisation of the acoustic chamber at NAL campus, a 16-ton shaker and a large (9 m) space simulation chamber with a sunbeam of about 4 m diameter through the efforts of Dr Chandramouli. Dr N K Mishra joined this group and contributed to further growth of ETF through systematic R&D on facilities, later setting up an integrated satellite integration facility at ISITE along with Mr T L Danabalan with very useful inputs from Mr V R Katti. This facility at ISITE is a workhorse today for producing 7–8 satellites a year and an integration cum testing complex which is unique. An added fall out of this facility is that a fully integrated satellite at ISITE is put into its container only when it is being shipped to the launch site.

## Small is Beautiful

With the development of integrated electronics into Application Specific Integrated Circuit (ASIC) and FPGAs in the nineties, the mass and power of satellite systems started coming down and hence the cost of making satellites and launching. This gave rise to many new players in the US and Europe to get into making small satellites that could be launched as piggyback with leftover weight after fulfilling the requirements of the main satellites at much lower costs of the order of \$1–\$10 million. Smaller, cheaper and beautiful became a buzz word and small satellites were proposed as a solution to everything in space. Professor Martin of Surrey University, UK, played an important role in popularising small satellites, helping many academic institutions across the world to make small satellites, mostly for taking earth images and bringing the thrill of being in space at affordable cost. University of Surrey also developed subsystems like star sensors and reaction wheels for small satellites.

Mr K Thyagarajan developed interest in leading the development of small satellites in ISAC. We felt that a small satellite program will become a platform to introduce advances in satellite technology and proving those new technologies at low cost and effort as a payload of opportunity on PSLV flights. Small satellite brings a lot of system engineering into practice as many functionalities have to be combined. The best example of system engineering is our mobile phone, combining digital communication, RF communication in multiple bands with state-of-the-art antennae, various sensors and application software from desktop, finance, data streaming, etc. To achieve even a small subset of such functionalities at a smaller cost of testing in space so that these could later be adopted in professional satellites with minimum risk was our motivation. Our first effort was to develop a bus management unit, a microprocessor-based system that could perform all digital functions like TC, TM, AOCS logics and control algorithms, digital filters, bus data communication, etc. The program did yield some interesting satellites like IMS1 and IMS2. Small satellites are becoming even smaller now with increased functionality.

As the technology of small satellites was evolving, it was not limited to only the use of micro-electronics and system integration but also the development of other related subsystems like MEMS-based sensors, gyros, accelerometers and actuators, using shape memory alloys and other smart materials, miniature reaction wheels, stars sensors, etc.

**Space Robotics**

Robotics is an area of interest from utility and technology development to imitate humans. Thoughts like refuelling of satellites in orbit, repairing/replacing defective parts, aged batteries in space or space tug to modify the orbit of a satellite have been going on since the early seventies. No satellite has yet been designed for on-orbit repair, but that will happen someday. Like on ground, a robot has to be designed for a specific purpose, but most technologies are common and space robotic activity started in the Mechanism group.

**Upagraha News Letter**

With so much of happenings across various disciplines, it was decided to bring a centre-level newsletter called UPAGRAH during the Aryabhata project itself. It covered major technical developments and any other news that general staff should know. Mr P N Srinivasan was the first editor of the newsletter. Over the years, Mr T K Jayaraman, Mr V R Katti and Dr (Mrs) Sheela Iyer functioned as editors with Mr Santosh Hebbar from PPEG providing support. The Upagraha continues even today and almost all centres of ISRO are now bringing their own in-house newsletters.

**Journal of Spacecraft Technology**

In spite of doing a lot of good R&D work, engineers at ISAC were shy of publishing their work in journals. To meet the aspirations of space technologists and encourage them to publish their contributions in a journal, in 1991, a journal of "Spacecraft Technology", a publication from ISRO Satellite Centre, was envisioned and the proposal was given by Mr A V Patki and Dr Surendra Pal. The first issue was out in August 1991. The editorial board consisted of Mr A V Patki as Chief Editor, Dr S Pal, Dr S Murugesan, Dr B V Sheela, Dr A K Sharma and Mr Santosh Hebbar. ISSN Number (0971–1600) was obtained. ISAC scientists were encouraged and trained to write technical papers on technologies developed in ISRO Satellite Centre. Since then, the journal is continuing its publications and became quite popular amongst ISAC Engineers. Its circulation has widened across the country and abroad.

**JCM**

Joint Consultative Machinery (JCM) is a very unique mechanism created by Prof Dhawan to bring harmony between working employees and the so-called management. JCMs are in each and every centre and there is one overall with the participation of all centres. The staff always come up with a long list of demands, but ISRO employees through evolution have been very understanding and considerate. Once explained why a particular demand cannot be met, they generally accept. But this goes by accepting those demands that can be accepted. This is the reason that we do not see discontent in ISRO and the staff continue to work with harmony and dedication.

# Chapter 8
# Indian Satellite Navigation Program

After operationalization of remote sensing and communication satellite programs, it was natural to think whether India should have its own Navigation Satellites.

Global Navigation Satellite Systems like GPS and GLONASS were already providing position, navigation and timing information, across the globe, but timing had selective availability. Only C/A code was available with GPS accuracy of around 50–100 m till May 2000.

GPS and GLONASS, although global systems, employed ionosphere and troposphere models which were applicable to the USA and parts of erstwhile USSR, respectively. In 1992, the USA started working on a Space Based Wide Area Augmentation System which was certified by Federal Aviation System (FAA) for Heli services, in January 2008. It was presumed, by most of the civil aviation authorities across the globe, that Global Navigation Satellite System (GNSS) consisting of GPS, GLONASS and other upcoming constellations will be part of civil aviation operations across the globe by 2020 to support civil aviation during partial visibility under foggy conditions.

International Civil Aviation Organisation (ICAO) started working in this direction and in its meeting in Canada and various players, Airport Authority of India (AAI) Director General of Civil Aviation (DGCA) and ISRO-SCPO representatives attended the regular meetings.

In the same duration, the European Union started working for having their own GALLELEO parallel to GPS and GLONASS constellations. They submitted a proposal in 1993 to MEA, Government of India, for having a collaborative and cooperative program related to commercial benefit of satellite navigation.

Almost during the same time, the French submitted a proposal to MEA, GoI, for inviting India to be a part of their program, with almost the cost of 600 million Euros. The European Union was developing its GALILEO program and Europe Geo Stationary Navigation Overlay System (EGNOS). With detailed discussion, it became clear that the participation was meant only in terms of financial contribution and not as part of constellation or support with security code as that was limited to Europe only. ISRO convinced MEA to decline the offer.

© Indian National Academy of Engineering 2022
P. S. Goel, *Making of a Satellite Centre*,
https://doi.org/10.1007/978-981-16-3480-2_8

## 8.1  GAGAN and IRNSS

Fog in winter season plays havoc with transportation in general and aviation in particular. I remember having spent 8 h at Delhi airport, sometime in the aircraft, waiting for the aircraft to take off. The concept of augmenting GPS for CAT II or III to assist aviation services was emerging in the US, and we thought it was even more important for India to have such an augmentation. We started dialogue with airport authority and they, though interested, were sceptic about certification process and wanted the US help or a replica of the US system. Even the US did not have a certified system at that time, though there was a very good knowledge base with industries like M/s Ratheon and academic institutions like MIT. A few experts of Indian origin in the US had developed expertise and were ready to help in developing the concept of augmentation. With this support, we decided to implement a satellite-based GPS augmentation system (S BAS) for supporting aviation in the Indian region and GPS Aided Geo Augmented Navigation (GAGAN) was born.

Dr S Pal headed a team consisting of Dr S V Kibe from ISRO HQ, Mr K N S Rao, Mr A S Ganesan, Mr S G Basu and others from ISAC, Mr Vilash Palsule, Mr A K Sisodia, Dr K S Dasgupta, Dr Sivaram and others from SAC, and Mr Soma, Mr Elango, Mr Rangararaju and Mr J Das from ISTRAC. A proposal was submitted to Chairman, ISRO, and later on a Memorandum of Understanding was signed between AAI and ISRO, for ISRO executing the project for the GEO Navigation Overlay System.

A Management team was constituted with Dr Pal as Chairman of Project Management Board with members from ISAC, DGCA and AAI. The team consisting of Dr Kibe from HQ, Mr K N S Rao (who was designated as Project Director), Mr Ganesan (Associate Project Director), Mr Soma from ISTRAC and Dr K S Dasgupta, Mr Sisodia and Mr Palsule from SAC, Chairman AAI and his representatives along with DGCA representatives were other members. The Project Management Board was chaired by Dr Pal, while the Project Management Council was chaired by Director, ISRO Satellite Centre.

It was decided that ISRO will provide the payload both from L1 to L5 frequency as a hosted payload, in three ISRO Communication Satellites. Ground segments will be procured. AAI will provide total real estate and other logistics for other ground segments. ISRO was to do handholding till the system becomes operational. A team of AAI engineers was stationed at ISRO Satellite Centre. After examining the global competitive bids for supplying ground segments, M/s Raytheon of the USA was selected as the prime contractor for the ground segment. The technology demonstration phase with the INMARSAT satellite was conducted during 2004–2007 period, while the final operation phase was conducted during 2008–2013. ISRO provided and hosted payload on GSAT 8, 10 and 15. ISRO also developed a dual-shell, grid-based ionosphere model for GAGAN over the Indian FIR region. The system became operational in 2015. The system consists of 3 satellites, 3 master control centre, 2 in Bangalore and one near Delhi, and 15 reference stations across the country all situated in AAI premises. Since 2015, the system has been handed over to AAI and ISRO is providing guidance and help.

Space debris has been a topic of interest to the space community, particularly dealing with regulations and international affairs. Mr A S Ganeshan in Flight Dynamics Division (FDD) was active in modelling aspects and in this regard was attending periodic meetings at international forums on behalf of ISRO. He used to brief me on return, requesting support and extra manpower. Having listened to him a few times, I realised that this pure academic exercise will not take him too far in ISRO except periodic foreign tours, though he was very good at modelling and analysis. Also, it was not contributing to ISRO or ISAC in any substantial manner.

We also realised that the GPS system was becoming inevitable for civil and military applications throughout the world. Even though there was a Presidential guarantee from the US not to disrupt the GPS civil support, P code was protected for the US and its military allies only. However, even NATO countries did not believe this declaration and Europe had started work on the alternate, Galileo system. Indian armed forces were very sceptic about the use of GPS for critical operations, were quite vocal but had no option. I asked Mr Ganeshan to study the possibility of a regional system using Geostationary and Geosynchronous satellites, giving similar accuracy, but minimising the number of satellites, to be controlled by Indian ground stations. We could not afford a global system of 28 satellites like GPS and our strategic needs are regional not global like that of the US. Mr Ganesan studied about 1800 cases along with Mr Rathnakar and Mr Rajneesh Gupta, Ms Nirmala and others. They came up with many alternatives and finally, we narrowed down to a seven-satellite configuration, three geostationary and four highly inclined geosynchronous satellites to serve the region of our strategic interest and the concept of IRNSS was born. Later, Mr Ganesan was so convinced of the concept and utility that he devoted full time to IRNSS only. But, the IRNSS is not a concept or constellation alone; it has to be an operational system with hardware, software and real-time connectivity of a very comprehensive network. We brought in a larger management team into operation with Mr K Suryanarana Rao as project director and Dr S Pal as Program Director as overall in charge.

For the IRNSS system, a constellation of 3 Geo Stationary satellites and 4 Geo Synchronous Satellites suitably distributed over Indian space and neighbourhood was chosen. Figure 8.1 shows the IRNSS constellation. A project presentation was made, under the chairmanship of Dr G Madhavan Nair to Space Commission and other relevant Government Departments/Authorities. Government of India approved the IRNSS project, on 9 May 2006. A Program Management Board was constituted under the Chairmanship of Dr Pal for the overall execution of the project, having Dr S V Kibe, ISRO HQ, Mr P Soma, ISTRAC, Mr K S Dasgupta, Mr Sisodia and Mr Vilash Palsule from SAC and Mr K N S Rao (Project Director) and Mr A S Ganesan (Associate Project Director) as members. The first IRNSS satellite was launched in 2013. The system was operational with 4 CDMA ranging stations across the country and with one master control, navigation and timing centre at Byalalu near Bangalore, with TTC uplinking station at MCF, Hassan, Bhopal, and ISTRAC, Lucknow. All the needed satellites are in orbit. Parallelly, SAC also developed Navigation Payloads

**Fig. 8.1** IRNS Constellation

under the leadership of Mr Nilesh Desai, besides developing IRNSS receivers and its applications.

At SAC, Ahmedabad, the development of an indigenous atomic clock has been undertaken. IRNSS has also got 16 reference stations for collecting data for iono-sphere, clock and orbital parameters of all satellites, which are sent to Mission Control Centre for uplinking the corrections and navigation messages. After the superannua-tion of Dr S Pal and Mr K N Rao, Mr A S Ganesan took over as Program Director and Project Director of IRNSS. Later on, Mr Ramasubramanian took over as Program Director, ISRO Satellite Navigation Program from Mr A S Ganesan. Needless to add here, Dr S V Kibe interacted with international and national agencies for coordina-tion related to frequency, code and other clearances. It was a gigantic task completed by the ISRO team in a limited time at a minimal cost. It is worth mentioning here that GAGAN became operational before EGNOS, and IRNSS constellation also got completed much before GALILEO.

IRNSS was dedicated to the fisherman community by Prime Minister Shri Narendra Modi and christened it as Navigation with Indian Constellation (NaviC).

The IRNSS team also carried out studies wherein with the addition of four more satellites in GSO orbits, NaviC's coverage, could extend from the existing 1500 kms around the GEO political boundary of India to provide coverage from Mongolia to Iraq and to further Eastern Europe with ±15 m positional accuracies.

It is to be noted that INRSS/NaviC is giving a similar level of Geo-positioning accuracy as the 28 satellite GPS system, in the specified region as defined.

## 8.2 The Space Shuttle and Solar Power Satellites

Yuri Gagarin's space flight in 1961, Neal Armstrong's moon landing in 1969 and the first geostationary satellite for communication in 1964 are some of the historical events in space. Equally and perhaps even more important was the development of the space shuttle. Space shuttle dominated the space news in the eighties and its aim was to provide quick and cheap access to space. The shuttle was a marvellous piece of engineering and technology, but an economic disaster. The three-engine main cryogenic stage with an external tank were unique technology achievements. Also, it had the two largest solid strap-on boosters. Rather than reducing the cost of launch by an order of magnitude (ten times less), the actual cost of maintaining a shuttle was bleeding NASA.

We understand that refurbishing each shuttle after every launch was costing NASA about $ 400 Million. Soviets also developed a shuttle "BURAN", which flew only once and was grounded for unknown reasons. This was an important message to the space community against hype that many a time has misled the space industry including solar power satellites. A solar panel of 11 sq km to be placed in orbit to generate 1GW of power and transfer it to the ground through microwave looked very simple in physics and many companies demonstrated power transfer from GEO. But the numbers tell the story another way. Even if the problems of maintaining an 11 SQ Km area panel towards the sun and 1 sq km dia antenna towards the earth is resolved, the shear launch cost of these materials to orbit will be 3–4 orders of magnitude higher than generating the same 1 GW power from solar panels on ground, though with the limitation of generating power for 8 h against 24 h from space. The launch cost numbers have not changed much for the last 50 years. What used to be about US $ 30,000 per kg to GTO in the 1970s has now come down to about US $ 12,000 per kg. This is not likely to vary much lower even for Space X's heavy booster (Big Falcon Rocket), perhaps lower by a factor of two or three. This may be good for going to Mars but not good enough for a solar power satellite in economic terms.

We adopted an approach of keeping our eyes open to the developments elsewhere, but chose our path very carefully so as not to deviate from the "man and society" theme. The logic has not changed for the last 50 years and is not likely to change for the next 50 years. Reusability of launch vehicle subsystem will definitely result in a reduction in the launch cost and it has to be the next direction for emerging launch vehicles of ISRO.

# Chapter 9
# From Technology to Management

I took over as Director ISAC on 1 November 1997 on the superannuation of Mr. Aravamudan. I went to Prof Dhawan to pay my respects and he just spoke one sentence, "**When you leave your desk in the evening, remember that someone else may occupy this desk tomorrow morning**". I never forgot this message in all my future roles, not just in ISAC but elsewhere also.

## 9.1 The Fast Track Satellite, INSAT-3B

Just a few months back, we had lost INSAT-2D satellite after 3 months of successful operation. ISAC was asked to prepare a proposal for procurement of a communication satellite by the SATCOM program office (ISRO HQ) to fill the gap. They had some offers to bring a satellite in 30 months. Mr Ramachandran had left ISRO for a job at INMARSAT and was located at Boeing Aerospace, Los Angeles, USA, and Mr V R Katti took over as Program Director, GEOSAT at ISAC. We gave a counter-proposal to build and launch our own fast track satellite in 30 months.

After prolonged discussions and accusations that we were delaying the telecom services to the User, we succeeded in stopping the import of a communication satellite. It was a matter of self-esteem for us in ISAC as well as for ISRO that after a successful INSAT-2 program and getting approval for the INSAT-3 series, we import a communication satellite. Simultaneously, we had the challenge of understanding the cause of the very sudden failure of a very professional satellite. We knew that it was a total power system failure, triggered by arcing in the satellite. Detailed data analysis and simulations carried by a team comprising Mr V R Katti, Mr Prahlad Rao, Dr S V K Shastry and Mr M Nageshwararao convinced us that it was a case of arcing and arc tracking near Solar Array Drive, and the basic cause was the Kapton wire that we were using almost everywhere. Detailed new guidelines for harness fabrication were prepared. It was a lesson at a huge cost, and many satellites of other countries had been lost due to such issues. We were lucky that we were not amongst the first few nations to get into GSO, but we had to learn fast from other's experiences.

© Indian National Academy of Engineering 2022
P. S. Goel, *Making of a Satellite Centre*,
https://doi.org/10.1007/978-981-16-3480-2_9

The fast track satellite was INSAT-3B and Dr P S Nair was pulled out from the Mechanical System Area to be appointed as Project Director, under GEOSAT Program. Dr Nair took this as his mission for life and was backed by the whole of the GEOSAT team at ISAC, SAC and other centres/units of ISRO. We launched INSAT-3B in March 2000. With INSAT-2C in service, there was no significant impact on INSAT services. INSAT-3B is another example of the "never say die" spirit of ISAC as the lead centre for satellites in India.

## 9.2   ISAC Council

ISAC council has been there for a long time but my experience of association for a few years was not so positive as most of the time was taken in discussing routine administrative matters and need aspect proposals. We decided to focus discussions on programs and policies and create separate mechanisms for routine administrative aspects. The council could only be briefed. One important observation I made was that though director ISAC chairs the LEOS council, there was an emotional gap between LEOS and ISAC and scientist-level interaction had got totally disrupted. Other than director ISAC, no other connection was existing and hence, I decided to have director LEOS (Dr Alex) as member of the ISAC council, as well as to have one of DD ISAC and Program Directors of IRS and GEOSAT be members of the LEOS council. This was helpful in bridging that gap and Dr Alex could smoothly take over as director ISAC after Dr Shankara who retired at a later date. We allocated certain time to discuss strategies, whether related to productionisation, technology development or future programs. This created a lot of discussions, sometimes perceived as a waste of time, but was important to develop future leadership.

## 9.3   TES

In the journey of our space program, the remaining relevant to the country came to a turning point with the country getting into the Kargil war in the year 1999. The war was thrust upon India by Pakistan by occupying advanced vacated army posts in Kargil at the end of winter, breaking the understanding of decades and taking strategic advantage of the height of the occupied posts. During May–June 1999, the brave Indian army was fighting a difficult war with heavy sacrifices being reported every day.

In mid-June, 1999, in the midst of the war, one evening, chairman Dr Kasturirangan walked into my room, unannounced. His 16.30 h flight to Delhi was cancelled by Air India and he was allotted a seat on the next flight at 19.50 h. Rather than going back to ISRO HQ, he decided to come to ISAC, being very close to HAL airport. Towards the end of our discussions, he asked "Can we do something for our soldiers

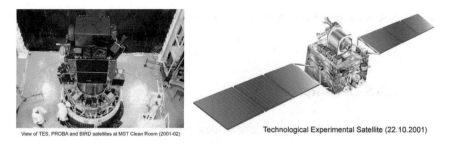

View of TES, PROBA and BIRD satellites at MST Clean Room (2001-02)

Technological Experimental Satellite (22.10.2001)

**Fig. 9.1**  TES in cleanroom

in Kargil?" So far, we had not thought of any security-related concern in ISRO. I just told him that I would come back within a week.

I called Dr George Joseph, director SAC, and he said, "Come tomorrow morning and let us discuss." I landed in Ahmedabad and had a day-long discussion with his team, Mr A S Kiran Kumar, Dr Ram Ratan and Mr Nagachenchaiah. By evening, we were ready with payload definition, a camera with 1 m resolution that can be built in 18 months, making use of electronics from IRS-1C, getting new optics, apparent velocity reduction by a factor of 5.8 through a "step and stare" imaging technique, ground station tracking phased array antenna for data dump and many other new technologies. We promised a schedule of 2 years for launch. Overall, we developed 13 new technologies and a new mission management of imaging satellite. The communication group under Dr Pal developed a 4 PSK data transmitter and a hemispherical phased array antenna in X band with dual-beam formation. In this concept, the two beams were tracking two ground stations, namely NRSA and Delhi simultaneously. This was an entirely new concept in the space domain. Totally new optics for the payload was designed and realised.

Dr Kasturirangan gave immediate oral approval to go ahead without project or funding sanction. Named Technology Experiment Satellite (TES), the satellite was launched on 22 October 2001. Figure 9.1 shows TES in clean room and in orbit configuration. Though the Kargil war lasted only a few weeks and TES did not contribute directly to the Kargil war, TES was an instant success with armed forces. It was so useful to them that we could not get any data from TES over the region of ISRO's interest, not even over a city like Bangalore or Delhi. But TES is not just a story of building a satellite in a very short time, it is the story of change of mindset, a story of extreme commitment to a cause, changed the perception of ISRO that defence space is not untouchable and many stories of personal sacrifices. A lady in ground checkout division called her mother-in-law to live with her, because she would not be able to take care of her children for the next 2 years. She was always there in the cleanroom whenever I visited. There were many others with such dedication. What was driving them? I had only made one statement, "If our satellite saves one of our soldier's life, then purpose of our life is served". What we achieved in TES as technical excellence and schedule was unprecedented in the country, perhaps in the world, only comparable to Aryabhata for schedule and much more in the overall national context.

## 9.4   METSAT (Kalpana)

It was well understood that the presence of meteorology payload VHRR was putting a lot of constraints on the configuration of INSAT satellites, making them inefficient. Mr V R Katti proposed to build an exclusive small satellite with a VHRR payload that can be launched on PSLV into GTO. Mission studies showed feasibility. However, a national committee under the chairmanship of Prof R Narasimha comprising of user ministry representatives and experts was constituted. Program Director GEOSAT Mr V R Katti represented ISAC. The committee just in one sitting approved the proposal and submitted its report to the chairman ISRO. Thus was the birth of METSAT and Mr V K Kaila was appointed as Project Director under GEOSAT Program. PSLV GTO capability was limited to about 1100 kg and hence the new spacecraft bus was named as I1K (representing 1 ton class INSAT). Being a small satellite, no sail and boom assembly were needed and only a large magnetic torque was sufficient to balance the radiation pressure, thus simplifying the satellite. Figure 9.2 shows METSAT in the cleanroom. It is just another example of innovative thinking and deviating from set ideas.

METSAT was a big success not only as a concept but also on an operational basis. Then Hon. Prime Minister Mr Atal Bihari Vajpayee named and dedicated the

**Fig. 9.2** METSAT in cleanroom

Kalpana (METSAT) - Integration (2001-02)

satellite after Ms Kalpana Chawla, the Space Shuttle lady astronaut of Indian origin who died along with co-astronauts while the space shuttle was re-entering into the atmosphere. It had a structural failure.

## 9.5  SRE

Yet another important development with much larger implication for ISRO is the Spacecraft Recovery Experiment (SRE), a joint mission between ISAC and VSSC. Developing re-entry technology was my dream and so was, as I discovered, of Dr B N Suresh, then associate director of VSSC, and we decided to take this project based on the respective strengths of our centres. It was unlike typical re-entry missions of short duration, but a long-term mission like a space laboratory and to bring the whole module back. ISAC had the experience of designing satellite systems for long life in orbit and VSSC with its strength in launch vehicles, in guidance, ceramic tiles to withstand high temperature and overall recovery in the open sea.

We depended on Arial Delivery Research and Development Establishment (ADRDE), a Defence Research and Development Organisation (DRDO) lab in Agra for design of the parachute system and they did a perfect job. Another DRDO lab in Chandigarh, Terminal Ballistics Research Laboratory, conducted high-speed tests on ground to validate some of our hypotheses and analysis. While it was a personally driven project by Dr Suresh and me, the actual SRE mission was conducted after I left ISRO for Department of Ocean Development as secretary in Delhi. Figure 9.3 shows the SRE recovery and re-entry schematic. It was a very successful mission, validating the technology in all respects though unfortunately, the SRE has never been repeated so far. The SRE experience was very useful in a successful hypersonic technology test. Today, we depend a lot on SRE data for our planning of the

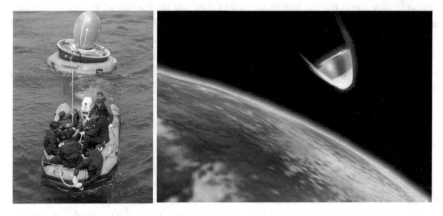

**Fig. 9.3**  SRE recovery and re-entry schematic

Gaganyaan mission. SRE was a well-conceived, well-planned and well-executed first (and last till Gaganyaan) joint project between ISAC and VSSC.

## 9.6   ISAC's Role as Lead Centre

Even as Associate Director of ISAC, I had this urge that as lead centre, we should be in the centre of planning the satellite technologies and program, but we were executing the ones planned by ISRO HQ. As director ISAC, I realised that the satellite program cannot be limited to Remote Sensing, Communication and Space Science, to be defined by program offices at ISRO Head Quarters as those program offices were lean and mostly busy in coordination with Delhi. Much more was needed in terms of technology and capability build-up through Technology Development Projects and conceiving/creating challenges for ourselves. METSAT, TES and SRE were the result of this thought process. We prepared our plan introducing new technologies through small satellites and other missions, embedding missions identified by the ISRO HQ. Professor Dhawan had initiated a very interesting exercise in the month of October every year as part of the budget exercise to have detailed presentations from all centres, units and projects/programs and outline the work to be undertaken by each in the next financial year. This continues even today in some form and is a forum to know what all is happening in ISRO and also to suggest new initiatives that are important for future. We used this forum to introduce the ISAC prepared plans.

## 9.7   TDPs

Realising that we had put basic satellites for remote sensing and satellite communication in place and the project mode R&D may not be the model for the future, we gave thrust to Technology Development Program (TDP), again an initiative of Prof Dhawan and Prof Rao. In addition to other milestone and related project reviews, I decided to conduct my own TDP review twice a year. We were far behind the state of the art as a lead centre for the satellite program, and TDP was the only mechanism to fill this gap.

## 9.8   Capacity Building

During 1997–98, we were making on an average one to two satellites in a year and as the demand was soaring, we decided to increase our capacity to 4–5 satellites per year and started working towards this goal. We had the capability but not the capacity. Productionisation required standardisation and common systems to the extent possible between INSAT and IRS programs along with the availability of components and materials. It was necessary to introduce outsourcing to industry for fabrication of as many subsystems as feasible and services like testing, environmental facility running and making use of other R&D labs like NAL for certain critical activities. We trained manpower in electronic/electrical manufacturing outsourced to industry and complete TM/TC package came through outsourcing. We qualified hybrid micro-packaging circuit fabrication in more than two private units and DC–DC converter fabrication in another private fabricator. It was important to expand our integration, ground check out and test facilities to commensurate with our plan to make 4–5 satellites per year by expanding the existing cleanrooms and planning new cleanrooms and other related facilities. We transferred technology for the fabrication of heat pipes to a new start-up company M/s Avasarala for production as we were importing these either from France or from Japan at a huge cost with long schedule implications. We transferred technology to produce C band TTC (Telemetry, Tracking and Command) transponders to SAMEER, Chennai, an autonomous body under Department of Electronics, GoI, for production and our aim was to make SAMEER our production partner and even an RF technology partner at a later date. We got a GPS receiver developed at M/s Accord, a private company in Bangalore, for satellite applications catering to high Doppler shift, specific to satellites. We organised training for industry to implement thermal control systems like fabrication of Multi-Layer Insulation (MLI) blankets. Our aim was to outsource as much routine activity to the industry as possible and to focus on new developments/technologies, not to do what others can do and cater to new and emerging challenges.

## 9.9    Satellite Production and Component Bank

While discussing how to increase the production of satellites, we realised that the availability of electronic components was a big handicap. Typically, procurement of components was done against each project, after the approval of the project. It took about 6 months to finalise the component list and later a minimum of another 8–9 months to place the order after negotiations. It was mostly minimum order quantity (MOQ) in the highest per piece slab. Delivery of class S space grade component might take 12–15 months. Hence, even a routine satellite project was taking about 4–5 years whereas foreign companies were doing the same in 30–36 months.

We decided to change our strategy and created component banks, one bank for remote sensing low earth-orbiting satellites and the other bank for geostationary satellites, for five satellites in each category, totalling components for 10 satellite models. This was also to reduce our administrative procurement efforts by a factor of ten, better quality control due to larger batch production and big saving in cost due to lower price for higher slabs. It was a bit difficult in the beginning to convince joint secretaries in the department and to allot a sum of about Rupees hundred crores for the component bank. However, the procurement exercise convinced everyone as we got quantity discounts of up to 40% (average discount being about 30%). It also required a more disciplined approach to component management, hence program offices were asked to create a cell to keep complete records in a computer database. The bonded store was totally reorganised with respect to environment control and ease of access. The cost of each project component was deducted from its budget and credited to the component bank.

## 9.10   Cultural Transformation

More than engaging the industry, it was a huge cultural transformation exercise at ISAC. Best engineers and technicians sincerely believed that what they were doing could not be done by anyone else in the country. We convinced individuals that it was a necessity to meet the increasing numbers and their own growth as ISRO demands something new every time from everyone.

The Central Electronics Fabrication Facility (CEFF) for fabricating electronics packages, for example, had developed expertise in soldering and assembly of digital, RF and power packages to the rigorous space quality. It takes on an average 3 years to qualify a wireman to do space grade soldering through theory and practical tests. They not only were (and are) doing excellent work, but they also believed that none else in the country could do soldering of printed circuit boards of satellites as they do. It took persistent effort to convince them to train and certify outsourced manpower. The first contract given to a private company was to make TM/TC digital package, and it took the company to create 31 production documents in almost 6 months covering minute details of the procedure, quality control and introducing regulatory mechanisms like material control board, waiver board, etc. ISAC was transforming, though, a bit slowly. This need for cultural transformation was not limited to ISAC alone but across all participating centres. Laboratory for Electro Optics Systems (LEOS) created a separate facility for industry producing Earth Sensors in old sheds vacated by ISAC.

I remember an interesting incident from earlier days. Dr Anantharam, Head, LPSC Bangalore, a very good friend of mine, was a very innovative person. He himself had developed a technique of loading catalyst in monopropellant thrusters and used to perform this task all by himself in secrecy. One day, I called him and asked how long you want to remain as a technician. He understood and trained a technician for catalyst loading but after a lot of persuasion. There are similar stories on the production of DTG and other elements. Cultural transformations are more difficult than solving hard technical problems, but we largely succeeded through persistent efforts and setting examples.

**Fig. 9.4** RISAT in
cleanroom

## 9.11   RISAT-1

We realised that microwave remote sensing/imaging was gaining importance particularly because the Indian sky is practically overcast during the Kharif season. Dr
Tapan Mishra from SAC made a proposal for a microwave Synthetic Aperture Radar
(SAR) using TR modules. The project looked too complex to be accepted at the first
glance, but seeing its importance to the agriculture in the Kharif season, we decided
to go ahead. Mr R N Tyagi was appointed as Project Director. There was a debate
on the selection of frequency band; L and S bands were considered good for agriculture, X band for defence application and C band, a kind of compromise between the
two. We, in consultation with many and on advice from Dr George Joseph, decided
to choose C band. This was a very new beginning for ISRO. RISAT-1 proved to
be another landmark mission in ISRO's remote sensing program. Figure 9.4 shows
RISAT-1 in cleanroom.

## 9.12   Cartosat-1

Encouraged by IRS-1C, there was a sudden spurt in remote sensing activity and
a demand came for cartography, 3D or Digital Elevation Modelling (DEM). It was
ideal to put three cameras, one forward-looking, second downward-looking and third,
backward-looking. However, it was not possible to accommodate three cameras in
one satellite with PSLV launch and we had to manage with two cameras. A study at
SAC led by Dr George Joseph indicated that a forward camera at 31 deg tilted about
pitch axis and second tilted backward by 5 Deg. would be the best option, meeting all
the mission objectives. For CARTO-1, the optics was large for 2.5 m resolution and
had to be procured from abroad. It was a signal to augment our facilities at LEOS.
CARTO-1 satellite again put India in the forefront of satellite technology, the best

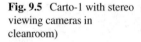
**Fig. 9.5**  Carto-1 with stereo viewing cameras in cleanroom)

cartography satellite in its category and its data is still the most used data for Digital Elevation Modelling. Figure 9.5 shows CARTO-1 in cleanroom.

## 9.13   New Campus for ISAC

I realised that the present campus at NAL, a mere 30-acre land, was highly inadequate for our dreams and felt that ISAC should have at least 200 acres of land. We also did not want to go much far away from Bangalore. When I presented this thought in one of the ISRO council meetings, the then joint secretary Mr Sengupta, had a hefty laugh, "Dr Goel wants 200 acres land in Bangalore for ISAC!".

I decided to pursue it on our own. In one of the meetings of Society for Aviation Technology Industry (SIATI), I had a talk with Dr Krishnadas Nair, the then Chairman HAL, who agreed to sell about 100 acres of land between Aeronautical Development Agency and the forthcoming new ring road. It was a dream come true and we purchased this 110 acres of land for Rs 95 crores. Dr Kasturirangan was very supportive and he himself negotiated this deal with Dr Krishnadas Nair. Today, it is known as ISITE and is a place of most of the action for ISAC. It will be impossible to think of URSC turning eight to ten satellites of different kinds without ISITE. We prepared a long-term road map (master plan) of ISITE, focusing on Indian industry participation, developing fabrication, integration and testing facilities for satellites up to six tonne. Our initial focus was to develop a cleanroom that could facilitate integrating a six-tonne satellite, a minimum of six satellites at a time, along with associated mechanical integration and test facilities, a 16-ton shaker, a 6 m thermo-vacuum chamber, EMI/EMC test facility, Compact Antenna Test Facility and an acoustic test facility, all arranged in such a way that the satellite need not have to be put in a container till it leaves for launch. This brilliant concept was developed and executed by Mr Dhanabalan and Dr N K Mishra with close interaction with Deputy Directors and Mr V R Katti, Program Director GEOSAT. Our master plan envisaged production and industry-related activities at ISITE and core R&D at the NAL campus. Today, ISITE is bubbling with activities, though with some differences to the original master plan, quite common with changing perceptions and priorities as time passes.

## 9.14  Chandrayaan 1

Like TES, there was one more change of direction in ISRO that happened around 1999–2000. The world space programs were moving towards outer space with Space Station taking shape, China and Japan showing much interest in mission to moon and Europe participating with the US and Russia both in multiple missions. Dr Kasturirangan posed a question, "having largely met the objectives set by the founder (Prof Sarabhai) of advanced technology for the benefit of common man, having one of the most comprehensive remote sensing program in the world encompassing the on board and ground segment, having a very robust communication satellite program and serving even the most demanding customer, INTELSAT through INSAT-2E and having Telemedicine and Tele education programs & services, should we not extend our vision beyond the founder's dream in 1967".

Professor Sarabhai had explicitly said that we have no fantasy of competing with advanced countries in exploring the Moon, but that was in 1967, some 30 odd years ago. Dr Kasturirangan suggested a nationwide debate on India going for the Moon mission. This was debated at multiple forums, viz., political, science and engineering academies, scientific think tanks and public platforms. Except for a few sceptics, there was overwhelming support to a mission to Moon and thus came the Chandrayaan-1 mission. It was the second major policy update, enhancing the ISRO vision of aspiring India.

Satellite technology is engineering-driven and Project Director of any mission so far had been an engineer who had developed some technology. Our mission team comes from a mixed background and even an engineer in the mission team really does not get into designing a system, though they have a good understanding of the technology. Mr Y N Bhushan working on the GEOSAT mission and heading the mission team had been requesting that there should be a chance for a mission person to become project director. Mr Annadurai had done a good job as mission manager for INSAT-2C, 2D and 2E missions. He had also contributed significantly as Associate Project Director of GSAT-3 (popularly known as EDUSAT). I realised that there was practically no new engineering development in Chandrayaan-1 as it

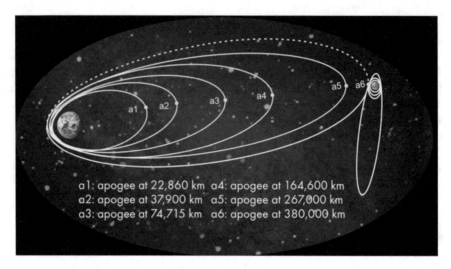

**Fig. 9.6** Chandrayaan Mission scenario

was to be built on the I1K platform of GEOSAT and the Chandrayaan I1K platform of GEOSAT was primarily a mission management challenge. Hence, we decided to make Mr Annadurai as Project Director of Chandrayaan-1. Figure 9.6 shows the Chandrayaan-1 mission scenario.

## 9.15    Lack of Linkages with the Society

I did often visit Prof U R Rao and Prof Dhawan to pay my respects and to brief them about progress, as their words were always inspiring. Towards 2000–2001, Prof Dhawan started to drop in once in a while and he was always concerned about society. Perhaps, you all know that he was a rationalist, not believing in religion or God, but he was deeply concerned about people, our people. He often told me that I was not doing enough for the society, why railway accidents? Why failed bridges? why so much corruption? all public-related issues. Once I gathered courage and said, "Sir, I am Director, ISAC and cannot interfere in these governance issues". His face became red, "Are you not an Indian?" That was a lesson and shook me to the core. Perhaps it changed me internally a bit, though not so visibly. I certainly understood Prof Dhawan a bit better.

## 9.16   Yet Another Change of Leadership

Mr Madhavan Nair took over as Chairman ISRO in July 2003 as Dr Kasturirangan was nominated to Rajya Sabha as MP, prior to the end of his extended term. It was a bit of sudden development though its planning had perhaps started much earlier in that year. Mr Madhavan Nair and I had worked together as directors, he being director of VSSC and LPSC, knew each other well, though poles apart in personality traits. We both were committed to ISRO and hence, we never had an occasion of difference of opinion to come up in our work as national and ISRO interests were beyond our personal, technical, temperamental and behavioural differences. I had sufficient freedom and autonomy with respect to the satellite program.

## 9.17   Widening the Landscape

We were busy with the INSAT-3 series, IRS series including CARTO-1, agile CARTOSAT series for the strategic program, SRE, ASTROSAT for astronomy, MEGHATROPIC, a joint CNES/ISRO mission for atmospheric sciences in addition to OCEANSAT, and Chandrayaan-1. Planning for the INSAT-4 series had also started. I had specifically got Mr Koteshwara Rao for ASTROSAT from LEOS and Dr Raju for Megha Tropique from SAC as Project Directors. We were writing a new script for the satellite program with help from SAC, LPSC Bangalore, IISU and the Composite group of VSSC. In addition, we were looking forward to making the ISRO program a truly national program through two routes. (1) Spread ISRO culture of working in project mode and outsourcing a lot of activity to industry including satellite integration and testing. We had engaged L&T for satellite integration of a communication satellite to start with and their team had been placed at ISAC. We had started giving subassembly work to industry, starting from solar panel hinges, solar cell laying on aluminium honeycomb substrate, heat pipes, TM/TC packages, RF transmitters and receivers, AOCE, etc. In LEOS also, a separate facility was created in old Aryabhata sheds to accommodate industry manpower to assemble and test Earth Sensors, sun sensors and magnetometers. (2) Creating quality culture in the Indian industry. We noticed that Maruti cars had done much more to bring quality mechanism to the automobile industry in the country and whether we in ISAC/ISRO could do the same to the Indian industry in general. Besides, the international community had now started taking ISRO seriously and willing to cooperate and partner with. I was interfacing Indo-US cooperation, for Chandrayaan-1 in particular.

## 9.18   Learning to Administrate

These are a few important noting from an administrative point of view and narration of a few examples will be useful to scientists getting into management positions. They also highlight some examples of ISAC culture.

(i)   I learned that there were undue delays in clearing payments after the items were delivered to stores. On scrutinising a few files, we discovered that in some cases, the indenter was not issuing an acceptance certificate for want of time for testing the equipment, not being aware of the financial implications of delays for the vendor in accepting the equipment. In some other cases, there were deliberate delays in making payments, being aware of financial implications when the market rate of interest was 15–18%. We set hard timelines for acceptance/rejection of the equipment and also for the release of payment after acceptance.

(ii)   There was a case of liquid Nitrogen payment in which our IFA found that the supplier was charging a higher rate for ISAC compared to SHAR though the contract term stipulated that the lowest price to any unit of ISRO would apply. The supplier requested to meet me alone to explain. A decision was taken never to meet any supplier/vender alone, without the presence of the concerned person(s). This avoids an unpleasant situation for any administrator.

(iii)   We used to have a very pleasant but frank relationship with our employees' union (Joint Consultative Machinery or JCM) and all differences were resolved with mutual understanding. We always gave our employees what we could and explained the reasons for what we could not. We had a very efficient and young IAS officer Mr Ambi at ISRO HQ at the level of director. He was deputed to assist Amarnath Yatra in 1998 and unfortunately he died in a landslide in the Himalayas. It was a terrible shock to ISRO and after about a few months of his death, I got an order from Mr Prabhakaran, then Additional Secretary with the approval of the Secretary DoS, to appoint Mr Ambi's wife on compassionate grounds. Before that, we had a pending list of 11 employees who died while working at ISAC in the past 5 years or so and no permission had come from ISRO HQ to appoint even a single eligible son or daughter, even after repeated reminders. On parameters (mostly economic) of compassionate appointment, Mrs Ambi came last. We could not have bypassed our own ISAC employees hence, no appointment was made. JCM members walked to my room and asked why I am not obeying the orders of ISRO HQ. I asked back why they were interested in the appointment of Mrs Ambi. They smiled and said that they wanted the management to make a mistake. "If you appoint Mrs Ambi, then we will force you to appoint all 11 pending compassionate appointments and you will have no choice". Administration is a very tight rope walking with no scope for committing a mistake. It is like a rocket and not a satellite.

(iv)   One day, Controller ISAC walked into my room and said that a newly recruited driver had complained that he had paid money for his appointment in ISAC. I said it was not possible because I had full confidence in the integrity of the selection committee chairman, Mr Srikanthan, a veteran and a sincere and honest person who came from APSU. On further investigation, the controller found that a committee member from ISAC used to note down addresses of candidates while scrutinising the applications and used to send a messenger to the house of selected candidates. The messenger would inform the candidate that he would get an appointment order, if he paid a fixed money as bribe, otherwise, he would not get the appointment order. Not knowing that his appointment order was already in the mail, he would pay the money for getting the job in the prestigious organisation. We had to move this gentleman to SHAR with the complete understanding with director SHAR. A police case would have created a sensation in the country, an ISRO scam, etc.

(v)    There was work pressure and a lot of persons were working after office hours (5.30 pm). We had the next dropping bus service at 7.30 pm and the overstaying staff would usually stay back for two extra hours. We were giving Tiffin at 6.00 pm against coupons. We were working on some Antrix projects and had savings in those accounts. We decided to use that money to serve evening Tiffin free, for the motivation of employees to work beyond office hours. A small adjustment for a good cause. I think the practice continues till today, hoping that an auditor will not read this noting.

(vi)   We found some very heartening stories of the dedication of our staff. While loading one of our satellite and test systems into the chartered aircraft in the late night to KUOROU for launch, our administration reported that labour had not reported for some reason, and the important cargo of about 15 lorries has to wait till morning. There was a possibility of rain too. While I was thinking of what to do next, our ISAC drivers came and asked my permission to load the cargo and when I permitted they did all the manual labour with care and efficiency. They said that these satellites give them the means to support their families, so what if they do manual labour work for a day.

(vii)  Similarly, a class four employee in the canteen was serving tea with such a pleasant smile and good manners and he was liked by everyone. He never went home if a meeting was going on irrespective of how late it was in the night. We wanted to recognise such persons and created a "Service Excellence Award". ISAC was the first centre in ISRO to introduce such an award and perhaps in the country. Though there are ample ways of awarding scientific staff, there was no method of awarding lower level staff for excellence in their service, which we called service with a smile, when the dedication comes from the heart. This award still continues in ISAC.

(viii) Once a chief minister of Karnataka called me over the phone and said that there was an interview the next day at ISAC for recruitment and that he had a person who was very good and should be selected. He mentioned some name. I explained to him that in ISRO/ISAC, the director appoints a committee and ensures that the committee makes recommendations based on the performance of the candidate only. My role was only to see that there was no other influence or criteria. I could not interfere in the process; otherwise, the whole of our value system will be distorted. He was very nice and said, "If this is your system, then let him compete". If we have conviction and adequately explain our stand, then even others would be supportive. Karnataka has always been supportive of S&T.

I wanted ISAC to be a role model in the country, innovative like Jet Propulsion Laboratory of NASA, leading in satellite technology for application-driven programs and achieving excellence in satellite engineering. ISRO should be part of the life of every Indian in day-to-day life. But, was my role over in ISAC? Perhaps "yes".

On 27 June 2005, I received an order from GoI, appointing me as Secretary Department of Ocean Development (DoD) from July 1st. It was really not as sudden as it looked. The story had started in March when during a Scientific Advisory Committee to the Cabinet (SAC-C) meeting in Vigyan Bhawan, Dr Harsh Gupta, then Secretary of DoD, asked me whether I would like to take over as Secretary DoD after his superannuation on 30 June 2005, and if yes he would recommend my name to the search and selection committee. I sought time and told him to come back within a week. Just 15 min later, during the same tea break, Prof Ramamurthy, the then Secretary, Department of Science and Technology (DST), asked me whether I would be interested in becoming Secretary DST on his superannuation on 31 May 2006, about a year and 3 months later, and again I requested for some time again. The thought of leaving ISAC/ISRO was something like separating body from the soul, but certain directions and decisions that ISRO had been taking were not very palatable to me. Will it be good for me? Will it be good for ISAC and ISRO? Will it be good for the country? I had no clear answer.

One thought that came to me was that perhaps my going out would be good for ISAC as my presence was a hindrance for HQ authority and hence owning it, so some bitterness towards ISAC. I consulted one of my important mentors, the only one. No, not Prof Rao. I knew that Prof Rao would never approve of my leaving ISAC, at any cost, personal or otherwise, as his soul lived in ISAC. The advice was that I could go as secretary DoD as the DST post was one more year away. There was also a suggestion to go to DST as Officer on Special Duty (OSD) till May 2006, but was not preferred. With uncertainty in Delhi, it is better not to wait that long. I was nearing 58.

A few weeks later, Prof Narasimha, another mentor, called me and said, " You will be offered Secretary DoD and please do not refuse. We want to create a Ministry of Earth Sciences (MoES) and you only can do it". He outlined the concept of MoES. It was an opportunity to do something new and unique for the country. I had decided that if the offer comes, I would take DoD and it was communicated to both, Dr Harsh Gupta and Prof Ramamurthy. However, all this was known to only very few and none in ISRO. It was a surprise to Mr Madhavan Nair, but he was supportive. Around the same time, Mr S K Das got empanelled as secretary and we both were felicitated together on June 29th. I was organising an Indo-US workshop till June 30th and my last assignment was over at 7.00 PM, in time for taking an early morning flight to Delhi. But that was not the end of my association with ISAC. In my farewell address on June 29th, I mentioned that my soul will wander around ISAC and will intervene whenever ISAC loses track. Unfortunately, my soul is still attached to my body and I could not live to that promise, but I am happy that ISAC has its own soul that corrects its path whenever there is a tendency to deviate from its destiny, a **"Centre of Excellence in satellite technology"**.

I requested Mr Madhavan Nair to bring Dr KN Shankara back to ISAC as Director as he had been Deputy Director, ESA, at ISAC and had done wonderful work as Director SAC. I was happy that Mr Nair agreed and that the legacy would continue. I met Prof Rao to take his blessings, a painful exercise for me as well as for him, but he understood with wet eyes and just said, "Best of luck". This perhaps was even more painful than my farewell address at ISAC.

# Chapter 10
# Looking from a Distance

As Secretary DoD and later MoES, I realised that ISRO to MoES is what NASA is to National Oceanic and Atmospheric Administration (NOAA) in the US and we need close coordination with ISRO for space segment for atmospheric sciences. VHRR in its evolution, Meghatropique satellite, new sensors like microwave sounders and multichannel imager for forthcoming satellites with committed schedule had to be coordinated and existing mechanism of National Natural Resource Management System (NNRMS) was invoked. NNRMS had practically ceased to exist. However, the committee for MoES met and tried to renew the system. One important request MoES made was to provide satellite connectivity of Indian Antarctic station to the mainland. The research team at Antarctic lived in isolation with very poor connectivity; a 3-minute call in 2 weeks with family was very stressing. Mr Madhavan Nair agreed and this connectivity has made a very significant contribution to the Antarctic program. Atmospheric program at SAC has been a major support to MoES in adopting satellite meteorology at India Meteorology Department (IMD) as IMD has not developed required R&D capability, largely focusing on operational aspects. This relationship with atmospheric science group at SAC has helped IMD in extracting atmospheric parameters from different satellite data and integrating into the numerical model for improved weather forecast.

With more and more exposure to working of other S&T ministries and departments while at Delhi, I realised a few important facts: (i) ISRO culture is unique and you do not find it elsewhere in the country. Delhi in particular being centre of politics has spoiled S&T culture. If a scientist is successful in Delhi, it is more likely that he has gone away from science and its culture (as well as ISRO culture) and has turned into a semi-politician. (ii) There is no organisation like ISRO for an individual scientist to be. The way ISRO nurtures young engineers through peer guidance, provides for very healthy competition, instils culture of hard work, recognises individual contribution and creates opportunities for bigger challenge makes ideal setting for Peter's principle to be applied. (iii) An ISRO person brought up with the kind of ISRO grill will succeed anywhere, in any circumstances. It is due to this training and ISRO culture that I succeeded in creating a new ministry, MoES, integrating atmospheric

© Indian National Academy of Engineering 2022
P. S. Goel, *Making of a Satellite Centre*,
https://doi.org/10.1007/978-981-16-3480-2_10

sciences, ocean sciences and technology and a few aspects of geology, in just 1 year. I retired as Secretary, MoES, on 30 April 2008, and Mr Madhavan Nair offered me Prof Vikram Sarabhai Distinguished Professorship at ISRO HQ Bangalore and I joined in July 2008.

While in Delhi as Secretary DOD, my information about ISRO activity was largely through newspapers. It was very pleasant to learn that SRE was a tremendous success. I was expecting a subsequent SRE mission that we had planned, but it never came. It is never possible to master technology like re-entry with just one mission. Other countries have repeated dozens of times. However, a similar mission, Pad Abort Test, came after about a decade and half later. Gaganyaan depends a lot on SRE engineering, developments, experience of recovery and the overall mission planning. The country did not recognise SRE for its technology achievements, but the country has never been properly appraised of technology achievements in science and technology by technology leaders. It is the media that decides the importance of a mission and the public perception is accordingly formed. It is a dangerous closed loop of public perception guiding the media and media nurturing the same perception. It will continue unless technical leadership takes special care and efforts.

An integrated satellite assembly and testing facility at ISITE came alive in 2007. One of the largest and tallest cleanroom buildings without a column in Asia, providing internal connectivity to 6 m thermo-vacuum chamber, vibration, acoustic testing and antenna Compact test facility, coordinated by Mr Danabalan, is unique in the world. I happened to see it a few years after its inauguration and it was a matter of great satisfaction. My visit was sometime after the visit of the then NASA chief who acknowledged that no facility like this exists in the United States. ISAC was becoming the pride of India and envy of even the most advanced nations. CARTOSAT-1 was another landmark mission, leaving a similar impact on the world's 3D mapping as IRS-1C had done for agriculture remote sensing.

I came to know that there were a number of reversals of policy. The L&T team was asked to leave citing that satellite integration cannot be outsourced. Long-term support to industry is a prerequisite for industry to enter the difficult space business as it requires excellence in engineering and quality but, low in numbers. This support was no longer available and in this place, emphasis was made to have more contract manpower at lower wages. Rather than strengthening remote sensing and communication program which had a lot of backlog, there was thrust only on one program, Chandrayaan-1. Seeing the impact of Chandrayaan-1, the thrust later shifted to Mangalyaan. There was not even one remote sensing satellite approval for 6 years. Obviously, it suited well, low-input, high-output missions with enormous visibility and political mileage.

ISAC had subsequently been relatively quick in seeing new directors. Dr Shankara retired in 2008 and Dr Alex came as director ISAC from LEOS. Dr Alex focused on Chandrayaan-1 and also tried to increase the ability to produce more satellites in line with the policy from ISRO HQ. He also tried to prepare ISAC for the human space program and went to acquire additional land in Yelahanka for setting up a separate unit. Dr K Radhakrishnan took over as Chairman ISRO in 2009. It is this time that so-called ISRO/DEVAS scam caught headlines. It was very painful to see ISRO in controversy and ISRO became the talk of the town for wrong reasons. It was a big setback to its image affecting the autonomy of ISRO. A low-level auditor started putting objections to the autonomy of ISRO centres citing General Financial Rules (GFR). This was a permanent setback, the procurement has slowed and authority to take decisions has been considerably diluted. Dr Alex tried to manage these events to keep going and boost the production of satellites. He superannuated in June 2012.

Mr S K Shiva Kumar was shifted as Associate Director ISAC from ISTRAC sometime earlier and took over as Director ISAC from Dr Alex. He was basically an ISACian, leading the mission team at ISAC before shifting to ISTRAC. Unfortunately, he developed health problems and could not focus as much as the job demands and carried with pursuing projects. Mr Shiv Kumar superannuated in March 2015.

Mr Kiran Kumar took over as Chairman ISRO in January 2015 in a difficult background of the so-called ISRO Scam and the pending list of projects, both in remote sensing and communication. His priority was to clear the backlog and restore the prestige of ISRO.

Dr Annadurai with his credentials as Project Director of Chandrayaan-1 took over as Director ISAC from Mr Shiva Kumar. He had acquired the image of a celebrity. He, however, put a lot of effort to augment satellite production capacity with hired technical manpower. This has been useful in the fast turnaround of satellites in recent times.

Dr Sivan took over as chairman in January 2018 from Mr A S Kiran Kumar. A veteran of launch vehicles, project director of GSLV Mark III, Director of LPSC and VSSC in the past, Dr Sivan is basically a project man like Prof UR Rao, works hard and drives harder, the projects, technology and the administration. There is some restoration of ISRO pride and confidence of technical leaders that files have started moving along the administrative channels without unnecessary queries to delay the files in the name of due diligence. Mr Kunhikrishnan, another veteran of launch vehicle and reputation of PSLV's most successful project director, has taken over as director ISAC in August 2018. ISRO is on rising path and ISAC has started searching for its new trajectory.

## 10.1   Space for National Security

It just so happened as a coincidence that three important events connecting me to national security happened within a period of just 1 week. While at ISRO HQ as Vikram Sarabhai Chair Professor in November 2008, I received a letter of membership of National Security Advisory Board (NSAB) on November 24th. Two days later, we had the 26/11 at Mumbai and on December 1st, I got the appointment letter as Chairman RAC/DRDO, Delhi, and I joined on 3 December 2008. All connected to National security in some way or the other. 26/11 had shaken every citizen of India and the 58 hour' ordeal made every technologist think why we are not able to provide any of the advanced technology tools to our armed forces while the terror bosses in Karachi were able to communicate with the terrorists in Mumbai in real time, making use of satellite phones. That is how our brave Sandeep was shot dead by those terrorists. It hurt me more, as I was also president of Indian National Academy of Engineering (INAE). Space for National Security became a topic of my focus.

We are the only country in the world where space technology was totally in the civil domain and we were not using it for National security even within the allowable domain of dual technology use. Whatever the use of space technology at that time, it was more of notional rather than any substantial value. The Satellite Based Surveillance (SBS) program of which I had some visibility was in the infant stage and limited to only spot imaging. Space is a great enabler, much more prior to the conflict than during the conflict, more strategic than tactical, more on preparing to be effective for a war rather than a direct means to win. There is a considerable confusion in the minds of planners of this capability, largely because all these planners are a product of the Internet, not even having seen a satellite from a distance and have no understanding of the capability of satellites in orbit. India needs another Sarabhai, not a scientist at PRL but a scientist in North Block. ISAC/URSC needs to reinvent to a new role, not directly as a doer but as an enabler for this great challenge. Like it is much easier to die for the country than to live for the country, it is easier to be a doer than to be an enabler.

In my first exposure to some of the DRDO laboratories as Chairman RAC, I noticed that DRDO had many technologies that were fairly well developed at Technology Readiness Level (TRL) of 6–8 to tackle insurgency at the local level as well as the terrorist's level, but not in product form that can be used by security agencies like National Security Guards. I, with the help of Principal Scientific Advisor (Dr Chidambaram) to the PM, organised a workshop in April 2019 and identified over a dozen such items. A new directorate of Low Intensity Conflict (LIC) was created at DRDO HQ for the same. It is unfortunate that none of these is still operational due to the lack of a champion.

My three-year term as Prof M G K Menon chair professor at RCI was essential to sow new seeds in the country, "Space for National Security". During a visit to DLRL, a laboratory at Hyderabad in 2010, I noticed a unique capability developed and being deployed at the border to identify, characterise and locate any radar within the line of sight. I asked a question, can we not put this in a satellite? Mr Bhupathi, director replied, "Sir, can you help". Thus started the project, Kautilya. I am happy to see that Kautilya launched in May 2019 has more than met its objectives and the seed has taken shape of a small plant. Dr Saraswat, then DG DRDO, and his successors Mr Avinash Chander and now Dr Satheesh Reddy have nurtured it with full conviction that Space for National Security needs to be taken to a more formal institutional commitment.

# Chapter 11
# The Journey Continues

I came back to Bangalore in May 2015 and Mr Kiran Kumar offered me Honorary Distinguished Professorship at ISRO HQ which I continue to occupy till now. It has given me another opportunity to be associated with ISRO and observe ISAC from not so much of a distance. Professor U R Rao passed away on 24 July 2017 after brief illness. He was coming to his office at ISRO HQ every day, busy with reading, writing and advising the current generation on various aspects of space science and engineering. He was Chairman of Space Science Advisory Committee and Chancellor of Indian Space Science and Technology Institute, Thiruvananthapuram, till his last day. As a mark of respect to this great son (scientist) of India and the founder of satellite technology, Government of India changed the name of his most beloved creation, ISRO Satellite Centre after his name as "U R Rao Satellite Centre" (URSC). Mr Kiran Kumar, then Chairman ISRO, needs to be thanked for this gesture.

URSC has transformed a lot in the last decade and half. SRE, Chandrayaan-1 and Mangalyaan have been landmark missions, making nationwide impact and bringing prestige to ISRO. It is producing about 8–9 satellites in the category of remote sensing, communication, navigation and space science and it is a big achievement. Chandrayaan-2 has been another landmark mission in technology and mission management, an effort that has put ISRO as the only shining star amongst the Indian science and technology institutions. Though the Lander failed to soft land at the designated spot, it hard landed very close to the designated site. We were very close but too far away. A complex interplay of disturbances and guidance algorithm during deboost and landing manoeuvres resulted in the last phase manoeuvres deviating a little from the designed plans. So many complex systems designed for orbiter and lander, so many new sensors and instrumentation working so precisely in the very first attempt are proofs that all participating agencies, not only team ISRO but even the participating industry, are now ready to undertake even greater challenges with confidence. URSC as lead centre in satellite technology has once again lives up to the dream of the founder, Prof Rao. Sad that he is not there to see it. He played a key a role in the selection of scientific payloads for Chandrayaan-2.

© Indian National Academy of Engineering 2022
P. S. Goel, *Making of a Satellite Centre*,
https://doi.org/10.1007/978-981-16-3480-2_11

But, the real challenge of "space for common man" is still far away. We still have hired satellite capacity for satellite communication in different categories. We still purchase a lot of remote sensing data from foreign satellites. Our NaviC is in the nascent stage of usage in civil and strategic sectors. Nevertheless, there is also a renewed commitment from ISRO to close these gaps in the next 3–4 years. Systematic planning has started and executive mechanism is being put in place.

Gaganyaan, another flagship program to put Indian men or women into space by the time country celebrates the 75th year of Independence in 2022, is yet another ambition of this aspiring nation. This is a huge task in terms of human rating of launch vehicle and the spacecraft, creating safety standards, certification mechanism, safe landing of the crew and building reliability with adequate redundancy and its management, are the important dimensions of this mission. This is a huge challenge for URSC and the newly created Human Space Flight Centre (HSFC).

However, the bigger challenge for URSC is to transform itself into a centre that provides leadership in various aspects of space for society. It has given two chairmen of ISRO in the past but fell short of giving its own leadership. URSC has to reinvent to nurture leadership in technology, create system engineers in large numbers, transform itself into an enabler to create capacity in the Indian industry and ensure that every citizen (a farmer, a fisherman, a businessman, a scientist, a person in hospital, a student or a soldier) feels enabled by space-based services. It is time to combine the visions of Prof Sarabhai, Prof Dhawan and Prof Rao and create URSC that other countries would like to take inspiration from, as we were trying to take inspiration from JPL. Can URSC reinvent itself once again? I believe it can.

**Homage, A few words about Prof Rao**

Professor U R Rao is a well-reported personality and you can get all information about him from numerous reports and books. I will only like to say a bit about the human in him and how he changed with time. An eldest son in a poor Brahmin family in a remote village near Udupi, he had to take care of relative's cattle to be able to go to school. As a student, he was brilliant and multifaceted, not just focusing on studies but also participating in all cultural activities. He went to BHU for his master's degree in physics, the first person not only in the family but in the entire village to get a master's degree in physics. As a researcher, he was very patient with analysing huge data, interpreting through mathematical tools and churning data into information. As in charge of a lab, he was focused on outcome and would go to any extent to pursue the end objective. He developed vision of a satellite centre for India that will build satellites to meet the vision of his Guru "Advanced technology to solve the problem of man and society" and develop applications for India to lead in space applications. He was a voracious reader, reading various books from science to fiction, management to finance. Once in a while play bridge (cards) and cut jokes under very serious situations.

Once in a meeting, defending erroneous working of his design an engineer said, "Sir, to err is human". Professor Rao's instantaneous remark was, "How many times you want to prove that you are a human being?" What was most remarkable was his quick analytical mind and decision-making capability. A very wide understanding of science and engineering, which probably came from his training as physicist, he could assess a very complex issue through a few questions in just a few seconds. His comments would be frank, sharp, witty and sometimes too harsh, but very simple and pure at heart, no hidden agenda and no polish in words. He was an institution in himself as the number of S&T leaders he developed in ISRO is incomparable to any other contemporary scientist barring Prof Sarabhai.

He was a nationalist to the core and never hesitated to express his views even when he knew that his boss (PM in most cases) has views to the contrary. Luckily, he had Prof Sarabhai and Prof Dhawan as technical bosses but he could express his difference of opinion even to political bosses as Secretary DoS. Towards the later years, he became more concerned about Indian society, perhaps an impact of his association with Prof Dhawan. He was the first person to talk about sustainable development at national and international forums and chaired committee of International Organisations and Developing Nations of International Astronautical Federation (Paris) for more than two decades. His versatility could be seen from the fact he was Chairman of Prasara Bharati Board and also a member of the Board of Reserve Bank of India, two different activities not even remotely connected to Science and Technology. But what we all miss is his concern and love for Space, ISAC (URSC) and his human touch. Long live Prof U R Rao.

Printed in the United States
by Baker & Taylor Publisher Services